POSSIBLE MINDS

BOOKS BY JOHN BROCKMAN

AS AUTHOR

By the Late John Brockman
37
Afterwords
The Third Culture
Digerati

AS EDITOR

About Bateson
Speculations
Doing Science
Ways of Knowing
Creativity
The Greatest Inventions of the Past 2,000 Years
The Next Fifty Years
The New Humanists
Curious Minds
What We Believe but Cannot Prove
My Einstein
Intelligent Thought
What Is Your Dangerous Idea?
What Are You Optimistic About?
Science at the Edge
What Have You Changed Your Mind About?
This Will Change Everything
Is the Internet Changing the Way You Think?
Culture
The Mind
This Will Make You Smarter
This Explains Everything
Thinking
What Should We Be Worried About?
The Universe
This Idea Must Die
What to Think About Machines That Think
Life
Know This
This Idea Is Brilliant

AS CO-EDITOR

How Things Are (with Katinka Matson)

POSSIBLE MINDS

Twenty-Five Ways of Looking at AI

Edited by

JOHN BROCKMAN

PENGUIN PRESS | NEW YORK | 2019

PENGUIN PRESS
An imprint of Penguin Random House LLC
penguinrandomhouse.com

LIBRARY OF CONGRESS CATALOGING-IN-PUBLICATION DATA

Names: Brockman, John, 1941- editor.
Title: Possible minds : twenty-five ways of looking at AI / edited by John Brockman.
Description: New York : Penguin Press, 2019. | Includes index.
Identifiers: LCCN 2018032888 | ISBN 9780525557999 (hardcover) |
ISBN 9780525558002 (ebook)
Subjects: LCSH: Artificial intelligence—Social aspects.
Classification: LCC Q335 .D436 2018 | DDC 006.3—dc23 LC record available at
https://lccn.loc.gov/2018032888

Printed in the United States of America
1 3 5 7 9 10 8 6 4 2

DESIGNED BY AMANDA DEWEY

For Einstein, Gertrude Stein, Wittgenstein, and Frankenstein

Acknowledgments

My thanks to Scott Moyers of Penguin Press for his editorial exuberance and my agent, Max Brockman, for his continued encouragement. A special thanks, once again, to Sara Lippincott for her thoughtful attention to the manuscript.

CONTENTS

CHAPTER 1. **Seth Lloyd: Wrong, but More Relevant Than Ever**   *1*

It is exactly in the extension of the cybernetic idea to human beings that Wiener's conceptions missed their target.

CHAPTER 2. **Judea Pearl: The Limitations of Opaque Learning Machines**   *13*

Deep learning has its own dynamics, it does its own repair and its own optimization, and it gives you the right results most of the time. But when it doesn't, you don't have a clue about what went wrong and what should be fixed.

CHAPTER 3. **Stuart Russell: The Purpose Put into the Machine**   *20*

We may face the prospect of superintelligent machines—their actions by definition unpredictable by us and their imperfectly specified objectives conflicting with our own—whose motivations to preserve their existence in order to achieve those objectives may be insuperable.

CHAPTER 4. **George Dyson: The Third Law**   *33*

Any system simple enough to be understandable will not be complicated enough to behave intelligently, while any system complicated enough to behave intelligently will be too complicated to understand.

INTRODUCTION:
ON THE PROMISE AND PERIL OF AI

A rtificial intelligence is today's story—the story behind all other stories. It is the Second Coming and the Apocalypse at the same time: good AI versus evil AI. This book comes out of an ongoing conversation with a number of important thinkers, both in the world of AI and beyond it, about what AI is and what it means. Called the Possible Minds Project, this conversation began in earnest in September 2016, in a meeting at the Grace Mayflower Inn & Spa in Washington, Connecticut, with some of the book's contributors.

What quickly emerged from that first meeting is that the excitement and fear in the wider culture surrounding AI now has an analog in the way Norbert Wiener's ideas regarding "cybernetics" worked their way through the culture, particularly in the 1960s, as artists began to incorporate thinking about new technologies into their work. I witnessed the impact of those ideas at close hand; indeed, it's not too much to say they set me off on my life's path. With the advent of the digital era beginning in the early 1970s, people stopped talking about Wiener, but today, his Cybernetic Idea has been so widely adopted that it's internalized to the point where it no longer needs a name. It's everywhere, it's in the air, and it's a fitting place to begin.

NEW TECHNOLOGIES = NEW PERCEPTIONS

Before AI, there was cybernetics—the idea of automatic, self-regulating control, laid out in Norbert Wiener's foundational text of 1948. I can date my own serious exposure to it to 1966, when the composer John Cage invited me and four or five other young arts people to join him for a series of dinners—an ongoing seminar about media, communications, art, music, and philosophy that focused on Cage's interest in the ideas of Wiener, Claude Shannon, and Marshall McLuhan, all of whom had currency in the New York art circles in which I was then moving. In particular, Cage had picked up on McLuhan's idea that by inventing electronic technologies we had externalized our central nervous system—that is, our minds—and that we now had to presume that "there's only one mind, the one we all share."

Ideas of this nature were beginning to be of great interest to the artists I was working with in New York at the Film-Makers' Cinematheque, where I was program manager for a series of multimedia productions called the New Cinema 1 (also known as the Expanded Cinema Festival), under the auspices of avant-garde filmmaker and impresario Jonas Mekas. They included visual artists Claes Oldenburg, Robert Rauschenberg, Andy Warhol, and Robert Whitman; kinetic artists Charlotte Moorman and Nam June Paik; happenings artists Allan Kaprow and Carolee Schneemann; dancer Trisha Brown; filmmakers Jack Smith, Stan Vanderbeek, Ed Emshwiller, and the Kuchar brothers; avant-garde dramatist Ken Dewey; poet Gerd Stern and the USCO group; minimalist musicians La Monte Young and Terry Riley; and, through Warhol, the music group The Velvet Underground. Many of these people were reading Wiener, and cybernetics was in the air. It was at one of these dinners that Cage reached into his briefcase and took out a copy of *Cybernetics* and handed it to me, saying, "This is for you."

During the festival, I received an unexpected phone call from Wiener's colleague Arthur K. Solomon, head of Harvard's graduate program in biophysics. Wiener had died the year before, and Solomon's and

Wiener's other close colleagues at MIT and Harvard had been reading about the Expanded Cinema Festival in the *New York Times* and were intrigued by the connection to Wiener's work. Solomon invited me to bring some of the artists up to Cambridge to meet with him and a group that included MIT sensory-communications researcher Walter Rosenblith, Harvard applied mathematician Anthony Oettinger, and MIT engineer Harold "Doc" Edgerton, inventor of the strobe light.

Like many other "art meets science" situations I've been involved in since, the two-day event was an informed failure: ships passing in the night. But I took it all on board and the event was consequential in some interesting ways—one of which came from the fact that they took us to see "the" computer. Computers were a rarity back then; at least none of us on the visit had ever seen one. We were ushered into a large space on the MIT campus, in the middle of which there was a "cold room" raised off the floor and enclosed in glass, in which technicians wearing white lab coats, scarves, and gloves were busy collating punch cards coming through an enormous machine. When I approached, the steam from my breath fogged up the window into the cold room. Wiping it off, I saw "the" computer. I fell in love.

Later, in the fall of 1967, I went to Menlo Park to spend time with Stewart Brand, whom I had met in New York in 1965 when he was a satellite member of the USCO group of artists. Now, with his wife, Lois, a mathematician, he was preparing the first edition of the *Whole Earth Catalog* for publication. While Lois and the team did the heavy lifting on the final mechanics for *WEC*, Stewart and I sat together in a corner for two days, reading, underlining, and annotating the same paperback copy of *Cybernetics* that Cage had handed to me the year before, and debating Wiener's ideas.

Inspired by this set of ideas, I began to develop a theme, a mantra of sorts, that has informed my endeavors since: "new technologies = new perceptions." Inspired by communications theorist Marshall McLuhan, architect-designer Buckminster Fuller, futurist John McHale, and cultural anthropologists Edward T. "Ned" Hall and Edmund Carpenter, I started reading avidly in the fields of information theory, cybernetics, and systems theory. McLuhan suggested I read biologist J. Z. Young's *Doubt and Certainty in Science,* in which he said that we create tools and we mold ourselves through our use of them. The other text he recommended was Warren Weaver and Claude Shannon's 1949 paper "Recent Contributions to the Mathematical Theory of Communication," which begins: "The word *communication* will be used here in a very broad sense to include all of the procedures by which one mind may affect another. This, of course, involves not only written and oral speech, but also music, the pictorial arts, the theater, the ballet, and in fact all human behavior."

Who knew that within two decades of that moment we would begin to recognize the brain as a computer? And in the next two decades, as we built our computers into the Internet, that we would begin to realize that the brain is not a computer but a network of computers? Certainly not Wiener, a specialist in analog feedback circuits designed to control machines, nor the artists, nor, least of all, myself.

"WE MUST CEASE TO KISS THE WHIP THAT LASHES US."

Two years after *Cybernetics,* in 1950, Norbert Wiener published *The Human Use of Human Beings*—a deeper story, in which he expressed his concerns about the runaway commercial exploitation and other unforeseen consequences of the new technologies of control. I didn't read *The Human Use of Human Beings* until the spring of 2016, when I picked up my copy, a first edition, which was sitting in my library next to *Cybernetics.* What shocked me was the realization of just how prescient Wiener was in 1950 about what's going on today. Although the first edition was a major bestseller—and, indeed, jump-started an important conversation—under pressure from his peers Wiener brought out a revised and milder edition in 1954, from which the original concluding chapter, "Voices of Rigidity," is conspicuously absent.

Science historian George Dyson points out that in this long-forgotten first edition, Wiener predicted the possibility of a "threatening new Fascism dependent on the *machine à gouverner*":

> No elite escaped his criticism, from the Marxists and the Jesuits ("all of Catholicism is indeed essentially a totalitarian religion") to the FBI ("our great merchant princes have looked upon the propaganda technique of the Russians, and have found that it is good") and the financiers lending their support "to make American capitalism and the fifth freedom of the businessman supreme throughout the world." Scientists . . . received the same scrutiny given the Church: "Indeed, the heads of great laboratories are very much like Bishops, with their association with the powerful in all walks of life, and the dangers they incur of the carnal sins of pride and of lust for power."

This jeremiad did not go well for Wiener. As Dyson puts it:

> These alarms were discounted at the time, not because Wiener was wrong about digital computing but because larger threats were looming as he completed his manuscript in the fall of 1949. Wiener had nothing against digital computing but was strongly opposed to nuclear weapons and refused to join those who were building digital computers to move forward on the thousand-times-more-powerful hydrogen bomb.

Since the original of *The Human Use of Human Beings* is now out of print, lost to us is Wiener's cri de coeur, more relevant today than when he wrote it sixty-eight years ago: "We must cease to kiss the whip that lashes us."

MIND, THINKING, INTELLIGENCE

Among the reasons we don't hear much about cybernetics today, two are central: First, although *The Human Use of Human Beings* was considered an important book in its time, it ran counter to the aspirations of many of Wiener's colleagues, including John von Neumann and Claude Shannon, who were interested in the commercialization of the new technologies. Second, computer pioneer John McCarthy disliked Wiener and refused to use Wiener's term "Cybernetics." McCarthy, in turn, coined the term "artificial intelligence" and became a founding father of that field.

As Judea Pearl, who, in the 1980s, introduced a new approach to artificial intelligence called Bayesian networks, explained to me:

> What Wiener created was excitement to believe that one day we are going to make an intelligent machine. He wasn't a computer scientist. He talked feedback, he talked communication, he talked analog. His working metaphor was a feedback circuit, which he was an expert in. By the time the digital age began in the early 1960s people wanted to talk programming, talk codes, talk about computational functions, talk about short-term memory, long-term memory—meaningful computer metaphors. Wiener wasn't part of that, and he didn't reach the new generation that germinated with his ideas. His metaphors were too old, passé. There were new means already available that were ready to capture the human imagination. By 1970, people were no longer talking about Wiener.

One critical factor missing in Wiener's vision was the cognitive element: mind, thinking, intelligence. As early as 1942, at the first of a series of foundational interdisciplinary meetings about the control of complex systems that would come to be known as the Macy Conferences, leading researchers were arguing for the inclusion of the cognitive element into the conversation. While von Neumann, Shannon, and Wiener were concerned about systems of control and communication of observed systems, Warren McCullough wanted to include mind. He turned to cultural anthropologists Gregory Bateson and Margaret Mead to make the connection to the social sciences. Bateson, in particular, was increasingly talking about patterns and processes, or "the pattern that connects." He called for a new kind of systems ecology in which organisms and the environment in which they live are one and the same and should be considered as a single circuit. By the early 1970s the cybernetics of observed systems—first-order cybernetics—moved to the cybernetics of observing systems—second-order cybernetics, or "the Cybernetics of Cybernetics," as coined by Heinz von Foerster, who joined the Macy Conferences in the mid-1950s and spearheaded the new movement.

Cybernetics, rather than disappearing, was becoming metabolized into *everything*, so we no longer saw it as a separate, distinct new discipline. And there it remains, hiding in plain sight.

"THE SHTICK OF THE STEINS"

My own writing about these issues at the time was on the radar screen of the second-order cybernetics crowd, including Heinz von Foerster as well as John Lilly and Alan Watts, who were the co-organizers of something called the AUM Conference, shorthand for "the American University of Masters," which took place in Big Sur in 1973, a gathering of philosophers, psychologists, and scientists, each of whom was asked to lecture on his own work in terms of its relationship to the ideas of British mathematician G. Spencer-Brown as presented in his book *Laws of Form*. I was a bit puzzled when I received an invitation—a very late invitation indeed—which they explained was based on their interest in the ideas I presented in a book called *Afterwords*, which were very much on their wavelength. I jumped at the opportunity, the main reason being that the keynote speaker was none other than Richard Feynman. I love to spend time with physicists, because they think about the universe, i.e., everything. And no physicist was reputed to be as articulate as Feynman. I couldn't wait to meet him. I accepted. That said, I am not a scientist, and I had never entertained the idea of getting on a stage and delivering a "lecture" of any kind, least of all a commentary on an obscure mathematical theory in front of a group identified as the world's most interesting thinkers. Only upon my arrival in Big Sur did I find out the reason for my very late invitation. "When is Feynman's talk?" I asked at the desk. "Oh, didn't Alan Watts tell you? Richard is ill and has been hospitalized. You're his replacement. And, by the way, what's the title of your keynote lecture?"

I tried to make myself invisible for several days. Alan Watts, realizing that I was avoiding the podium, woke me up one night with a three a.m. knock on the door of my room. I opened the door to find him standing in front of me wearing a monk's robe with a hood covering much of his face. His arms extended, he held a lantern in one hand and a magnum of scotch in the other. "John," he said in a deep voice with a rich aristo-cratic British accent, "you are a phony. And, John," he continued, "I am a phony. But, John, I am a *real* phony!"

The next day I gave my lecture, titled "Einstein, Gertrude Stein, Wittgenstein, and Frankenstein." Einstein: the revolution in 20th cen-tury physics. Gertrude Stein: the first writer who made integral to her work the idea of an indeterminate and discontinuous universe. Words represented neither character nor activity: A rose is a rose is a rose, and a universe is a universe is a universe. Wittgenstein: the world as limits of language. "The limits of my language mean the limits of my world." The end of the distinction between observer and observed. Frankenstein: cybernetics, AI, robotics, all the essayists in this volume.

The lecture had unanticipated consequences. Among the partici-pants at the AUM Conference were several authors of number one *New York Times* bestsellers, yet no one there had a literary agent. And I real-ized that all were engaged in writing a genre of book both unnamed and unrecognized by New York publishers. Since I had an MBA from Co-lumbia Business School and a series of relative successes in business, I was dragooned into becoming an agent, initially for Gregory Bateson and John Lilly, whose books I sold quickly, and for sums that caught my attention, thus kick-starting my career as a literary agent.

I never did meet Richard Feynman.

THE LONG AI WINTERS

This new career put me in close touch with most of the AI pioneers, and over the decades I rode with them on waves of enthusiasm, and into valleys of disappointment. In the early eighties the Japanese government mounted a national effort to advance AI. They called it the Fifth Generation; their goal was to change the architecture of computation by breaking "the von Neumann bottleneck" by creating a massively parallel computer. In so doing, they hoped to jump-start their economy and become a dominant world power in the field. In 1983, the leader of the Japanese Fifth Generation consortium came to New York for a meeting organized by Heinz Pagels, the president of the New York Academy of Sciences. I had a seat at the table alongside the leaders of the first generation, Marvin Minsky and John McCarthy; the second generation, Edward Feigenbaum and Roger Schank; and Joseph Traub, head of the National Supercomputer Consortium.

In 1981, with Heinz's help, I had founded The Reality Club (the precursor to the nonprofit Edge.org), whose initial interdisciplinary meetings took place in the boardroom at the NYAS. Heinz was working on his book *The Dreams of Reason: The Computer and the Rise of the Science of Complexity*, which he considered to be a research agenda for science in the 1990s.

Through the Reality Club meetings, I got to know two young researchers who were about to play key roles in revolutionizing computer science. At MIT in the late seventies, Danny Hillis developed the algorithms that made possible the massively parallel computer. In 1983, his company, Thinking Machines, built the world's fastest supercomputer by utilizing parallel architecture. His "connection machine" closely reflected the workings of the human mind. Seth Lloyd at Rockefeller University was undertaking seminal work in the fields of quantum computation and quantum communications, including proposing the first technologically feasible design for a quantum computer.

And the Japanese? Their foray into artificial intelligence failed and was followed by twenty years of anemic economic growth. But the leading U.S. scientists took this program very seriously. And Feigenbaum, who was the cutting-edge computer scientist of the day, teamed up with Pamela McCorduck to write a book on these developments. *The Fifth Generation: Artificial Intelligence and Japan's Computer Challenge to the World* was published in 1983. We had a code name for the project: "It's coming, it's coming!" But it didn't come; it went.

From that point on I've worked with researchers in nearly every variety of AI and complexity, including Rodney Brooks, Hans Moravec, John Archibald Wheeler, Benoit Mandelbrot, John Henry Holland, Danny Hillis, Freeman Dyson, Chris Langton, J. Doyne Farmer, Geoffrey West, Stuart Russell, and Judea Pearl.

AN ONGOING DYNAMICAL EMERGENT SYSTEM

From the initial meeting in Washington, Connecticut, to the present, I arranged a number of dinners and discussions in London and Cambridge, Massachusetts, as well as a public event at London's City Hall. Among the attendees were distinguished scientists, science historians, and communications theorists, all of whom have been thinking seriously about AI issues for their entire careers.

I commissioned essays from a wide range of contributors, with or without references to Wiener (leaving it up to each participant). In the end, twenty-five people wrote essays, all individuals concerned about what is happening today in the age of AI. *Possible Minds* is not *my* book, rather it is *our* book: Seth Lloyd, Judea Pearl, Stuart Russell, George Dyson, Daniel C. Dennett, Rodney Brooks, Frank Wilczek, Max Tegmark, Jaan Tallinn, Steven Pinker, David Deutsch, Tom Griffiths, Anca Dragan, Chris Anderson, David Kaiser, Neil Gershenfeld, W. Daniel Hillis, Venki Ramakrishnan, Alex "Sandy" Pentland, Hans Ulrich Obrist, Alison Gopnik, Peter Galison, George M. Church, Caroline A. Jones, and Stephen Wolfram.

I see the Possible Minds Project as an ongoing dynamical emergent system, a presentation of the ideas of a community of sophisticated thinkers who are bringing their experience and erudition to bear in challenging the prevailing digital AI narrative as they communicate their thoughts to one another. The aim is to present a mosaic of views that will help make sense out of this rapidly emerging field.

I asked the essayists to consider:

a. The Zen-like poem "Thirteen Ways of Looking at a Blackbird" by Wallace Stevens, which he insisted was "not meant to be a collection of epigrams or of ideas, but of sensations." It is an exercise in "perspectivism," consisting of short, separate sections, each of which mentions blackbirds in some way. The poem is about his own imagination; it concerns what he attends to.

b. The parable of the blind men and an elephant. Like the elephant, AI is too big a topic for any one perspective, never mind the fact that no two people seem to see things the same way.

What do we want the book to do? Stewart Brand has noted that "revisiting pioneer thinking is perpetually useful. And it gives a long perspective that invites thinking in decades and centuries about the subject. All contemporary discussion is bound to age badly and immediately without the longer perspective."

Danny Hillis wants people in AI to realize how they've been programmed by Wiener's book. "You're executing its road map," he says, "and you just don't realize it."

Dan Dennett would like to "let Wiener emerge as the ghost at the banquet. Think of it as a source of hybrid vigor, a source of unsettling ideas to shake up the established mind-set."

Neil Gershenfeld argues that "stealth remedial education for the people running the 'Big Five' would be a great output from the book."

Freeman Dyson, one of the few people alive who knew Wiener, notes that "*The Human Use of Human Beings* is one of the best books ever written. Wiener got almost everything right. I will be interested to see what your bunch of wizards will do with it."

THE EVOLVING AI NARRATIVE

Things have changed—and they remain the same. Now AI is everywhere. We have the Internet. We have our smartphones. The founders of the dominant companies—the companies that hold "the whip that lashes us"—have net worths of $65 billion, $90 billion, $130 billion. High-profile individuals such as Elon Musk, Nick Bostrom, Martin Rees, Eliezer Yudkowsky, and the late Stephen Hawking have issued dire warnings about AI, resulting in the ascendancy of well-funded institutes tasked with promoting "Nice AI." But will we, as a species, be able to control a fully realized, unsupervised, self-improving AI? Wiener's warnings and admonitions in *The Human Use of Human Beings* are now very real, and they need to be looked at anew by researchers at the forefront of the AI revolution. Here is Dyson again:

> Wiener became increasingly disenchanted with the "gadget worshipers" whose corporate selfishness brought "motives to automatization that go beyond a legitimate curiosity and are sinful in themselves." He knew the danger was not machines becoming more like humans but humans being treated like machines. "The world of the future will be an ever more demanding struggle against the limitations of our intelligence," he warned in *God & Golem, Inc.,* published in 1964, the year of his death, "not a comfortable hammock in which we can lie down to be waited upon by our robot slaves."

It's time to examine the evolving AI narrative by identifying the leading members of that mainstream community along with the dissidents and presenting their counternarratives in their own voices.

The essays that follow thus constitute a much-needed update from the field.

—*John Brockman*
New York, 2019

Chapter 1

WRONG, BUT MORE RELEVANT THAN EVER

SETH LLOYD

Seth Lloyd is a theoretical physicist at MIT, Nam P. Suh Professor in the Department of Mechanical Engineering, and an external professor at the Santa Fe Institute.

I met **Seth Lloyd** in the late 1980s, when new ways of thinking were everywhere: the importance of biological organizing principles, the computational view of mathematics and physical processes, the emphasis on parallel networks, the importance of nonlinear dynamics, the new understanding of chaos, connectionist ideas, neural networks, and parallel distributive processing. The advances in computation during that period provided us with a new way of thinking about knowledge.

Seth likes to refer to himself as a quantum mechanic. He is internationally known for his work in the field of quantum computation, which attempts to harness the exotic properties of quantum theory, like superposition and entanglement, to solve problems that would take several lifetimes to solve on classical computers.

In the essay that follows, he traces the history of information theory from Norbert Wiener's prophetic insights to the predictions of a technological "singularity" that some would have us believe will supplant the human species. His takeaway on the recent programming method known as deep learning is to call for a more modest set of expectations; he notes that despite AI's enormous advances, robots "still can't tie their own shoes."

It's difficult for me to talk about Seth without referencing his relationship with his friend and professor, the late theoretical physicist Heinz Pagels of Rockefeller University. The graduate student and the professor each had a profound effect on the other's ideas.

In the summer of 1988, I visited Heinz and Seth at the Aspen Center for Physics. Their joint work on the subject of complexity was featured in the current issue of *Scientific American*; they were ebullient. That was just two weeks before Heinz's tragic death in a hiking accident while descending Pyramid Peak with Seth. They were talking about quantum computing.

The Human Use of Human Beings, Norbert Wiener's 1950 popularization of his highly influential book Cybernetics: or Control and Communication in the Animal and the Machine (1948), investigates the interplay between human beings and machines in a world in which machines are becoming ever more computationally capable and powerful. It is a remarkably prescient book, and remarkably wrong. Written at the height of the Cold War, it contains a chilling reminder of the dangers of totalitarian organizations and societies, and of the danger to democracy when it tries to combat totalitarianism with totalitarianism's own weapons.

Wiener's Cybernetics looked in close scientific detail at the process of control via feedback. ("Cybernetics," from the ancient Greek for "helmsman," is the etymological basis of our word "governor," which is what James Watt called his pathbreaking feedback control device that transformed the use of steam engines.) Because he was immersed in problems of control, Wiener saw the world as a set of complex, interlocking feedback loops, in which sensors, signals, and actuators such as engines interact via an intricate exchange of signals and information. The engineering applications of Cybernetics were tremendously influential and effective, giving rise to rockets, robots, automated assembly lines, and a host of precision-engineering techniques—in other words, to the basis of contemporary industrial society.

Wiener had greater ambitions for cybernetic concepts, however, and in *The Human Use of Human Beings* he spells out his thoughts on its application to topics as diverse as Maxwell's Demon, human language, the brain, insect metabolism, the legal system, the role of technological innovation in government, and religion. These broader applications of cybernetics were an almost unequivocal failure. Vigorously hyped from the late 1940s to the early 1960s—to a degree similar to the hype of computer and communication technology that led to the dot-com crash of 2000–2001—cybernetics delivered satellites and telephone switching systems but generated few if any useful developments in social organization and society at large.

Nearly seventy years later, however, *The Human Use of Human Beings* has more to teach us humans than it did the first time around. Perhaps the most remarkable feature of the book is that it introduces a large number of topics concerning human/machine interactions that are still of considerable relevance. Dark in tone, the book makes several predictions about disasters to come in the second half of the 20th century, many of which are almost identical to predictions made today about the second half of the 21st.

For example, Wiener foresaw a moment in the near future of 1950 in which humans would cede control of society to a cybernetic artificial intelligence, which would then proceed to wreak havoc on humankind. The automation of manufacturing, Wiener predicted, would both create large advances in productivity and displace many workers from their jobs—a sequence of events that did indeed come to pass in the ensuing decades. Unless society could find productive occupations for these displaced workers, Wiener warned, revolt would ensue.

But Wiener failed to foresee crucial technological developments. Like pretty much all technologists of the 1950s, he failed to predict the computer revolution. Computers, he thought, would eventually fall in price from hundreds of thousands of (1950s) dollars to tens of thousands; neither he nor his compeers anticipated the tremendous explosion of computer power that would follow the development of the transistor and the integrated circuit. Finally, because of his emphasis on control, Wiener could not foresee a technological world in which innovation and self-organization bubble up from the bottom rather than being imposed from the top.

Focusing on the evils of totalitarianism (political, scientific, and religious), Wiener saw the world in a deeply pessimistic light. His book warned of the catastrophe that awaited us if we didn't mend our ways, fast. The current world of human beings and machines, more than a half century after its publication, is much more complex, richer, and contains a much wider variety of political, social, and scientific systems than he was able to envisage. The warnings of what will happen if we get it wrong, however—for example, control of the entire Internet by a global totalitarian regime—remain as relevant and pressing today as they were in 1950.

WHAT WIENER GOT RIGHT

Wiener's most famous mathematical works focused on problems of signal analysis and the effects of noise. During World War II, he developed techniques for aiming antiaircraft fire by making models that could predict the future trajectory of an airplane by extrapolating from its past behavior. In *Cybernetics* and in *The Human Use of Human Beings*, Wiener notes that this past behavior includes quirks and habits of the human pilot, thus a mechanized device can predict the behavior of humans. Like Alan Turing, whose Turing Test suggested that computing machines could give responses to questions that were indistinguishable from human responses, Wiener was fascinated by the notion of capturing human behavior by mathematical description. In the 1940s, he applied his knowledge of control and feedback loops to neuromuscular feedback in living systems, and was responsible for bringing Warren McCulloch and Walter Pitts to MIT, where they did their pioneering work on artificial neural networks.

Wiener's central insight was that the world should be understood in terms of information. Complex systems, such as organisms, brains, and human societies, consist of interlocking feedback loops in which signals exchanged between subsystems result in complex but stable behavior. When feedback loops break down, the system goes unstable. He constructed a compelling picture of how complex biological systems function, a picture that is by and large universally accepted today.

Wiener's vision of information as the central quantity in governing the behavior of complex systems was remarkable at the time. Nowadays, when cars and refrigerators are jammed with microprocessors and much of human society revolves around computers and cell phones connected by the Internet, it seems prosaic to emphasize the centrality of information, computation, and communication. In Wiener's time, however, the first digital computers had only just come into existence, and the Internet was not even a twinkle in the technologist's eye.

Wiener's powerful conception of not just engineered complex systems but all complex systems as revolving around cycles of signals and computation led to tremendous contributions to the development of complex human-made systems. The methods he and others developed for the control of missiles, for example, were later put to work in building the Saturn V moon rocket, one of the crowning engineering achievements of the 20th century. In particular, Wiener's applications of cybernetic concepts to the brain and to computerized perception are the direct precursors of today's neural-network-based deep-learning circuits, and of artificial intelligence itself. But current developments in these fields have diverged from his vision, and their future development may well affect the human uses both of human beings and of machines.

WHAT WIENER GOT WRONG

It is exactly in the extension of the cybernetic idea to human beings that Wiener's conceptions missed their target. Setting aside his ruminations on language, law, and human society for the moment, look at a humbler but potentially useful innovation that he thought was imminent in 1950. Wiener notes that prosthetic limbs would be much more effective if their wearers could communicate directly with their prosthetics by their own neural signals, receiving information about pressure and position from the limb and directing its subsequent motion. This turned out to be a much harder problem than Wiener envisaged: Seventy years down the road, prosthetic limbs that incorporate neural feedback are still in the very early stages. Wiener's concept was an excellent one—it's just that the problem of interfacing neural signals with mechanical-electrical devices is hard.

More significantly, Wiener (along with pretty much everyone else in 1950) greatly underappreciated the potential of digital computation. As noted, Wiener's mathematical contributions were to the analysis of signals and noise and his analytic methods apply to continuously varying, or analog, signals. Although he participated in the wartime development of digital computation, he never foresaw the exponential explosion of computing power brought on by the introduction and progressive miniaturization of semiconductor circuits. This is hardly Wiener's fault: The transistor hadn't been invented yet, and the vacuum-tube technology of the digital computers he was familiar with was clunky, unreliable, and unscalable to ever larger devices. In an appendix to the 1948 edition of *Cybernetics*, he anticipates chess-playing computers and predicts that they'll be able to look two or three moves ahead. He might have been surprised to learn that within half a century a computer would beat the human world champion at chess.

TECHNOLOGICAL OVERESTIMATION AND THE EXISTENTIAL RISKS OF THE SINGULARITY

When Wiener wrote his books, a significant example of technological overestimation was about to occur. The 1950s saw the first efforts at developing artificial intelligence, by researchers such as Herbert Simon, John McCarthy, and Marvin Minsky, who began to program computers to perform simple tasks and to construct rudimentary robots. The success of these initial efforts inspired Simon to declare that "machines will be capable, within twenty years, of doing any work a man can do." Such predictions turned out to be spectacularly wrong. As they became more powerful, computers got better and better at playing chess because they could systematically generate and evaluate a vast selection of possible future moves. But the majority of predictions of AI, e.g., robotic maids, turned out to be illusory. When Deep Blue beat Garry Kasparov at chess in 1997, the most powerful room-cleaning robot was a Roomba, which moved around vacuuming at random and squeaked when it got caught under the couch.

Technological prediction is particularly chancy, given that technologies progress by a series of refinements, halted by obstacles and overcome by innovation. Many obstacles and some innovations can be anticipated, but more cannot. In my own work with experimentalists on building quantum computers, I typically find that some of the technological steps I expect to be easy turn out to be impossible, whereas some of the tasks I imagine to be impossible turn out to be easy. You don't know until you try.

In the 1950s, partly inspired by conversations with Wiener, John von Neumann introduced the notion of the "technological singularity." Technologies tend to improve exponentially, doubling in power or sensitivity over some interval of time. (For example, since 1950, computer technologies have been doubling in power roughly every two years, an observation enshrined as Moore's Law.) Von Neumann extrapolated from the observed exponential rate of technological improvement to predict that "technological progress will become incomprehensively rapid and complicated," outstripping human capabilities in the not too distant future. Indeed, if one extrapolates the growth of raw computing power—expressed in terms of bits and bit flips—into the future at its current rate, computers should match human brains sometime in the next two to four decades (depending on how one estimates the information-processing power of human brains).

The failure of the initial overly optimistic predictions of AI dampened talk about the technological singularity for a few decades, but since the 2005 publication of Ray Kurzweil's *The Singularity IS Near,* the idea of technological advance leading to superintelligence is back in force. Some believers, Kurzweil included, regard this singularity as an opportunity: Humans can merge their brains with the superintelligence and thereby live forever. Others, such as Stephen Hawking and Elon Musk, worried that this superintelligence would prove to be malign and regarded it as the greatest existing threat to human civilization. Still others, including some of the contributors to the present volume, think such talk is overblown.

Wiener's lifework and his failure to predict its consequences are intimately bound up in the idea of an impending technological singularity. His work on neuroscience and his initial support of McCulloch and Pitts adumbrated the startlingly effective deep-learning methods of the present day. Over the past decade, and particularly in the last five years, such deep-learning techniques have finally exhibited what Wiener liked to call *Gestalt*—for example, the ability to recognize that a circle is a circle even if when slanted sideways it looks like an ellipse. His work on control, combined with his work on neuromuscular feedback, was significant for the development of robotics and is the inspiration for neural-based human/machine interfaces. His lapses in technological prediction, however, suggest that we should take the notion of a technological singularity with a grain of salt. The general difficulties of technological prediction and the problems specific to the development of a superintelligence should warn us against overestimating both the power and the efficacy of information processing.

THE ARGUMENTS FOR SINGULARITY SKEPTICISM

No exponential increase lasts forever. An atomic explosion grows exponentially, but only until it runs out of fuel. Similarly, the exponential advances in Moore's Law are starting to run into limits imposed by basic physics. The clock speed of computers maxed out at a few gigahertz a decade and a half ago, simply because the chips were starting to melt. The miniaturization of transistors is already running into quantum-mechanical problems due to tunneling and leakage currents. Eventually, the various exponential improvements in memory and processing driven by Moore's Law will grind to a halt. A few more decades, however, will probably be time enough for the raw information-processing power of computers to match that of brains—at least by the crude measures of number of bits and number of bit-flips per second.

Human brains are intricately constructed, the process of millions of years of natural selection. In Wiener's time, our understanding of the architecture of the brain was rudimentary and simplistic. Since then, increasingly sensitive instrumentation and imaging techniques have shown our brains to be far more varied in structure and complex in function than Wiener could have imagined. I recently asked Tomaso Poggio, one of the pioneers of modern neuroscience, whether he was worried that computers, with their rapidly increasing processing power, would soon emulate the functioning of the human brain. "Not a chance," he replied.

The recent advances in deep learning and neuromorphic computation are very good at reproducing a particular aspect of human intelligence focused on the operation of the brain's cortex, where patterns are processed and recognized. These advances have enabled a computer to beat the world champion not just of chess but of Go, an impressive feat, but they're far short of enabling a computerized robot to tidy a room. (In fact, robots with anything approaching human capability in a broad range of flexible movements are still far away—search "robots falling down." Robots are good at making precision welds on assembly lines, but they still can't tie their own shoes.)

Raw information-processing power does not mean sophisticated information-processing power. While computer power has advanced exponentially, the programs by which computers operate have often failed to advance at all. One of the primary responses of software companies to increased processing power is to add "useful" features, which often make the software harder to use. Microsoft Word reached its apex in 1995 and has been slowly sinking under the weight of added features ever since. Once Moore's Law starts slowing down, software developers will be confronted with hard choices between efficiency, speed, and functionality.

A major fear of the singulariteers is that as computers become more involved in designing their own software they'll rapidly bootstrap themselves into achieving superhuman computational ability. But the evidence of machine learning points in the opposite direction. As machines become more powerful and capable of learning, they learn more and more as human beings do—from multiple examples, often under the supervision of human and machine teachers. Education is as hard and slow for computers as it is for teenagers. Consequently, systems based on deep learning are becoming more rather than less human. The skills they bring to learning are not "better than" but "complementary to" human learning: Computer learning systems can identify patterns that humans cannot—and vice versa. The world's best chess players are neither computers nor humans but humans working together with computers. Cyberspace is indeed inhabited by harmful programs, but these primarily take the form of malware—viruses notable for their malign mindlessness, not for their superintelligence.

WHITHER WIENER

Wiener noted that exponential technological progress is a relatively modern phenomenon and not all of it is good. He regarded atomic weapons and the development of missiles with nuclear warheads as a recipe for the suicide of the human species. He compared the headlong exploitation of the planet's resources with the Mad Tea Party of *Alice in Wonderland*: Having laid waste to one local environment, we make progress simply by moving on to lay waste to the next. Wiener's optimism about the development of computers and neuromechanical systems was tempered by his pessimism about their exploitation by authoritarian governments, such as the Soviet Union, and the tendency for democracies, such as the United States, to become more authoritarian themselves in confronting the threat of authoritarianism.

What would Wiener think of the current human use of human beings? He would be amazed by the power of computers and the Internet.
He would be happy that the early neural nets in which he played a role
have spawned powerful deep-learning systems that exhibit the perceptual ability he demanded of them—although he might not be impressed
that one of the most prominent examples of such computerized *Gestalt*
is the ability to recognize photos of kittens on the World Wide Web.
Rather than regarding machine intelligence as a threat, I suspect he
would regard it as a phenomenon in its own right, different from and
co-evolving with our own human intelligence.

Unsurprised by global warming—the Mad Tea Party of our era—
Wiener would applaud the exponential improvement in alternative-energy
technologies and would apply his cybernetic expertise to developing the
intricate set of feedback loops needed to incorporate such technologies
into the coming smart electrical grid. Nonetheless, recognizing that the
solution to the problem of climate change is at least as much political as it
is technological, he would undoubtedly be pessimistic about our chances
of solving this civilization-threatening problem in time. Wiener hated
hucksters—political hucksters most of all—but he acknowledged that
hucksters would always be with us.

It's easy to forget just how scary Wiener's world was. The United
States and the Soviet Union were in a full-out arms race, building hydro-
gen bombs mounted on nuclear warheads carried by intercontinental
ballistic missiles guided by navigation systems to which Wiener
himself—to his dismay—had contributed. I was four years old when
Wiener died. In 1964, my nursery school class was practicing duck and
cover under our desks to prepare for a nuclear attack. Given the human
use of human beings in his own day, if he could see our current state,
Wiener's first response would be to be relieved that we are still alive.

Chapter 2

THE LIMITATIONS OF OPAQUE LEARNING MACHINES

JUDEA PEARL

Judea Pearl *is a professor of computer science and director
of the Cognitive Systems Laboratory at UCLA. His most recent book,
co-authored with Dana Mackenzie, is* The Book of Why:
The New Science of Cause and Effect.

In the 1980s, **Judea Pearl** introduced a new approach to artificial intelligence called Bayesian networks. This probability-based model of machine reasoning enabled machines to function—in a complex and uncertain world—as "evidence engines," continuously revising their beliefs in light of new evidence.

Within a few years, Judea's Bayesian networks had completely overshadowed the previous rule-based approaches to artificial intelligence. The advent of deep learning—in which computers, in effect, teach themselves to be smarter by observing tons of data—has given him pause, because this method lacks transparency.

While recognizing the impressive achievements in deep learning by colleagues such as Michael I. Jordan and Geoffrey Hinton, he feels uncomfortable with this kind of opacity. He set out to understand the theoretical limitations of deep-learning systems and points out that basic barriers exist that will prevent them from achieving a human kind of intelligence, no matter what we do. Leveraging the computational benefits of Bayesian networks, Judea realized that the combination of simple graphical models and data could also be used to represent and infer cause-effect relationships. The significance of this discovery far transcends its roots in artificial intelligence. His latest book explains causal thinking to the general public; you might say it is a primer on how to think even though human.

Judea's principled, mathematical approach to causality is a profound contribution to the realm of ideas. It has already benefited virtually every field of inquiry, especially the data-intensive health and social sciences.

As a former physicist, I was extremely interested in cybernetics. Though it did not utilize the full power of Turing Machines, it was highly transparent, perhaps because it was founded on classical control theory and information theory. We are losing this transparency now, with the deep-learning style of machine learning. It is fundamentally a curve-fitting exercise that adjusts weights in intermediate layers of a long input-output chain.

I find many users who say that it "works well and we don't know why." Once you unleash it on large data, deep learning has its own dynamics, it does its own repair and its own optimization, and it gives you the right results most of the time. But when it doesn't, you don't have a clue about what went wrong and what should be fixed. In particular, you do not know if the fault is in the program, in the method, or because things have changed in the environment. We should be aiming at a different kind of transparency.

Some argue that transparency is not really needed. We don't understand the neural architecture of the human brain, yet it runs well, so we forgive our meager understanding and use human helpers to great advantage. In the same way, they argue, why not unleash deep-learning systems and create intelligence without understanding how they work? I buy this argument to some extent. I personally don't like opacity, so I won't spend my time on deep learning, but I know that it has a place in the makeup of intelligence. I know that nontransparent systems can do marvelous jobs, and our brain is proof of that marvel.

But this argument has its limitations. The reason we can forgive our meager understanding of how human brains work is because our brains work the same way, and that enables us to communicate with other humans, learn from them, instruct them, and motivate them in our own native language. If our robots will all be as opaque as AlphaGo, we won't be able to hold a meaningful conversation with them, and that would be unfortunate. We will need to retrain them whenever we make a slight change in the task or in the operating environment.

So rather than experimenting with opaque learning machines, I am trying to understand their theoretical limitations and examine how these limitations can be overcome. I do it in the context of causal-reasoning tasks, which govern much of how scientists think about the world and, at the same time, are rich in intuition and toy examples, so we can monitor the progress in our analysis. In this context, we've discovered that some basic barriers exist, and that unless they are breached we won't get a real human kind of intelligence no matter what we do. I believe that charting these barriers may be no less important than banging our heads against them.

Current machine-learning systems operate almost exclusively in a statistical, or model-blind, mode, which is analogous in many ways to fitting a function to a cloud of data points. Such systems cannot reason about "What if?" questions and, therefore, cannot serve as the basis for Strong AI—that is, artificial intelligence that emulates human-level reasoning and competence. To achieve human-level intelligence, learning machines need the guidance of a blueprint of reality, a model—similar to a road map that guides us in driving through an unfamiliar city.

To be more specific, current learning machines improve their performance by optimizing parameters for a stream of sensory inputs received from the environment. It is a slow process, analogous to the natural-selection process that drives Darwinian evolution. It explains how species like eagles and snakes have developed superb vision systems over millions of years. It cannot explain, however, the super-evolutionary process that enabled humans to build eyeglasses and telescopes over barely a thousand years. What humans had that other species lacked was a mental representation of their environment—a representation that they could manipulate at will to imagine alternative hypothetical environments for planning and learning.

Historians of *Homo sapiens* such as Yuval Noah Harari and Steven Mithen are in general agreement that the decisive ingredient that gave our ancestors the ability to achieve global dominion about forty thousand years ago was their ability to create and store a mental representation of their environment, interrogate that representation, distort it by mental acts of imagination, and finally answer the "What if?" kinds of questions. Examples are interventional questions ("What if I do such-and-such?") and retrospective or counterfactual questions ("What if I had acted differently?"). No learning machine in operation today can answer such questions. Moreover, most learning machines do not possess a representation from which the answers to such questions can be derived.

With regard to causal reasoning, we find that you can do very little with any form of model-blind curve fitting, or any statistical inference, no matter how sophisticated the fitting process is. We have also found a theoretical framework for organizing such limitations, which forms a hierarchy.

On the first level, you have statistical reasoning, which can tell you only how seeing one event would change your belief about another. For example, what can a symptom tell you about a disease?

Then you have a second level, which entails the first but not vice versa. It deals with actions. "What will happen if we raise prices?" "What if you make me laugh?" That second level of the hierarchy requires information about interventions that is not available in the first. This information can be encoded in a graphical model, which merely tells us which variable responds to another.

The third level of the hierarchy is the counterfactual. This is the language used by scientists. "What if the object were twice as heavy?" "What if I were to do things differently?" "Was it the aspirin that cured my headache or the nap I took?" Counterfactuals are at the top level in the sense that they cannot be derived even if we could predict the effects of all actions. They need an extra ingredient, in the form of equations, to tell us how variables respond to changes in other variables.

One of the crowning achievements of causal-inference research has been the algorithmization of both interventions and counterfactuals, the top two layers of the hierarchy. In other words, once we encode our scientific knowledge in a model (which may be qualitative), algorithms exist that examine the model and determine if a given query, be it about an intervention or about a counterfactual, can be estimated from the available data—and, if so, how. This capability has dramatically transformed the way scientists are doing science, especially in such data-intensive sciences as sociology and epidemiology, for which causal models have become a second language. These disciplines view their linguistic transformation as the Causal Revolution. As Harvard social scientist Gary King puts it, "More has been learned about causal inference in the last few decades than the sum total of everything that had been learned about it in all prior recorded history."

As I contemplate the success of machine learning and try to extrapolate it to the future of AI, I ask myself, Are we aware of the basic limitations that were discovered in the causal-inference arena? Are we prepared to circumvent the theoretical impediments that prevent us from going from one level of the hierarchy to another level?

I view machine learning as a tool to get us from data to probabilities. But then we still have to make two extra steps to go from probabilities into real understanding—two big steps. One is to predict the effect of actions, and the second is counterfactual imagination. We cannot claim to understand reality unless we make the last two steps.

In his insightful book *Foresight and Understanding* (1961), the philosopher Stephen Toulmin identified the transparency-versus-opacity contrast as the key to understanding the ancient rivalry between Greek and Babylonian sciences. According to Toulmin, the Babylonian astronomers were masters of black-box predictions, far surpassing their Greek rivals in accuracy and consistency of celestial observations. Yet science favored the creative-speculative strategy of the Greek astronomers, which was wild with metaphorical imagery: circular tubes full of fire, small holes through which celestial fire was visible as stars, and hemispherical Earth riding on turtleback. It was this wild modeling strategy, not Babylonian extrapolation, that jolted Eratosthenes (276–194 BC) to perform one of the most creative experiments in the ancient world and calculate the circumference of the Earth. Such an experiment would never have occurred to a Babylonian data fitter.

Model-blind approaches impose intrinsic limitations on the cognitive tasks that Strong AI can perform. My general conclusion is that human-level AI cannot emerge solely from model-blind learning machines; it requires the symbiotic collaboration of data and models.

Data science is a science only to the extent that it facilitates the interpretation of data—a two-body problem, connecting data to reality. Data alone are hardly a science, no matter how "big" they get and how skillfully they are manipulated. Opaque learning systems may get us to Babylon, but not to Athens.

Chapter 3

THE PURPOSE PUT INTO
THE MACHINE

STUART RUSSELL

Stuart Russell is a professor of computer science and Smith-Zadeh Professor in Engineering at UC Berkeley. He is the co-author (with Peter Norvig) of Artificial Intelligence: A Modern Approach.

Computer scientist **Stuart Russell**, along with Elon Musk, Stephen Hawking, Max Tegmark, and numerous others, has insisted that attention be paid to the potential dangers in creating an intelligence on the superhuman (or even the human) level—an AGI, or artificial general intelligence, whose programmed purposes may not necessarily align with our own.

His early work was on understanding the notion of "bounded optimality" as a formal definition of intelligence that you can work on. He developed the technique of rational metareasoning, "which is, roughly speaking, that you do the computations that you expect to improve the quality of your ultimate decision as quickly as possible." He has also worked on the unification of probability theory and first-order logic—resulting in a new and far more effective monitoring system for the Comprehensive Nuclear Test Ban Treaty—and on the problem of decision making over long timescales (his presentations on the latter topic are usually titled "Life: play and win in 20 trillion moves").

He is very concerned with the continuing development of autonomous weapons, such as lethal microdrones, which are potentially scalable into weapons of mass destruction. He drafted the letter from forty of the world's leading AI researchers to President Obama that resulted in high-level national-security meetings.

His current work centers on the creation of what he calls "provably beneficial" AI. He wants to ensure AI safety by "imbuing systems with explicit uncertainty" about the objectives of their human programmers, an approach that would amount to a fairly radical reordering of current AI research.

Stuart is also on the radar of anyone who has taken a course in computer science in the last twenty-odd years. He is co-author of "the" definitive AI textbook, with an estimated 5-million-plus English-language readers.

Among the many issues raised in Norbert Wiener's *The Human Use of Human Beings* (1950) that are currently relevant, the most significant to the AI researcher is the possibility that humanity may cede control over its destiny to machines.

Wiener considered the machines of the near future as far too limited to exert global control, imagining instead that machines and machine-like control systems would be wielded by human elites to reduce the great mass of humanity to the status of "cogs and levers and rods." Looking further ahead, he pointed to the difficulty of correctly specifying objectives for highly capable machines, noting

> a few of the simpler and more obvious truths of life, such as that when a djinnee is found in a bottle, it had better be left there; that the fisherman who craves a boon from heaven too many times on behalf of his wife will end up exactly where he started; that if you are given three wishes, you must be very careful what you wish for.

The dangers are clear enough:

> Woe to us if we let [the machine] decide our conduct, unless we have previously examined the laws of its action, and know fully that its conduct will be carried out on principles acceptable to us! On the

other hand, the machine like the djinnee, which can learn and can make decisions on the basis of its learning, will in no way be obliged to make such decisions as we should have made, or will be acceptable to us.

Ten years later, after seeing Arthur Samuel's checker-playing program learn to play checkers far better than its creator, Wiener published "Some Moral and Technical Consequences of Automation" in *Science*. In this paper, the message is even clearer:

> If we use, to achieve our purposes, a mechanical agency with whose operation we cannot efficiently interfere . . . we had better be quite sure that the purpose put into the machine is the purpose which we really desire.

In my view, this is the source of the existential risk from superintelligent AI cited in recent years by such observers as Elon Musk, Bill Gates, Stephen Hawking, and Nick Bostrom.

PUTTING PURPOSES INTO MACHINES

The goal of AI research has been to understand the principles underlying intelligent behavior and to build those principles into machines that can then exhibit such behavior. In the 1960s and 1970s, the prevailing theoretical notion of intelligence was the capacity for logical reasoning, including the ability to derive plans of action guaranteed to achieve a specified goal. More recently, a consensus has emerged around the idea of a rational agent that perceives, and acts in order to maximize, its expected utility. Subfields such as logical planning, robotics, and natural-language understanding are special cases of the general paradigm. AI has incorporated probability theory to handle uncertainty, utility theory to define objectives, and statistical learning to allow machines to adapt to new circumstances. These developments have created strong connections to other disciplines that build on similar concepts, includ-ing control theory, economics, operations research, and statistics.

In both the logical-planning and rational-agent views of AI, the ma-chine's objective—whether in the form of a goal, a utility function, or a reward function (as in reinforcement learning)—is specified exoge-nously. In Wiener's words, this is "the purpose put into the machine." Indeed, it has been one of the tenets of the field that AI systems should be general purpose—i.e., capable of accepting a purpose as input and then achieving it—rather than special purpose, with their goal implicit in their design. For example, a self-driving car should accept a destina-tion as input instead of having one fixed destination. However, some aspects of the car's "driving purpose" are fixed, such as that it shouldn't hit pedestrians. This is built directly into the car's steering algorithms rather than being explicit: No self-driving car in existence today "knows" that pedestrians prefer not to be run over.

Putting a purpose into a machine that optimizes its behavior accord-ing to clearly defined algorithms seems an admirable approach to ensur-ing that the machine's "conduct will be carried out on principles acceptable to us!" But, as Wiener warns, we need to put in the right purpose. We might call this the King Midas problem: Midas got exactly what he asked for—namely, that everything he touched would turn to gold—but too late he discovered the drawbacks of drinking liquid gold and eating solid gold. The technical term for putting in the right pur-pose is *value alignment*. When it fails, we may inadvertently imbue ma-chines with objectives counter to our own. Tasked with finding a cure for cancer as fast as possible, an AI system might elect to use the entire human population as guinea pigs for its experiments. Asked to de-acidify the oceans, it might use up all the oxygen in the atmosphere as a side effect. This is a common characteristic of systems that optimize: Vari-ables not included in the objective may be set to extreme values to help optimize that objective.

Unfortunately, neither AI nor other disciplines (economics, statistics, control theory, operations research) built around the optimization of ob-jectives have much to say about how to identify the purposes "we really desire." Instead, they assume that objectives are simply implanted into the machine. AI research, in its present form, studies the ability to achieve objectives, not the design of those objectives.

Steve Omohundro has pointed to a further difficulty, observing that intelligent entities must act to preserve their own existence. This tendency has nothing to do with a self-preservation instinct or any other biological notion; it's just that an entity cannot achieve its objectives if it's dead. According to Omohundro's argument, a superintelligent machine that has an off switch—which some, including Alan Turing himself, in a 1951 talk on BBC Radio 3, have seen as our potential salvation—will take steps to disable the switch in some way.* Thus we may face the prospect of superintelligent machines—their actions by definition unpredictable by us and their imperfectly specified objectives conflicting with our own—whose motivations to preserve their existence in order to achieve those objectives may be insuperable.

1001 REASONS TO PAY NO ATTENTION

Objections have been raised to these arguments, primarily by researchers within the AI community. The objections reflect a natural defensive reaction, coupled perhaps with a lack of imagination about what a superintelligent machine could do. None hold water on closer examination. Here are some of the more common ones:

- *Don't worry, we can just switch it off.*† This is often the first thing that pops into a layperson's head when considering risks from superintelligent AI—as if a superintelligent entity would never think of that. This is rather like saying that the risk of losing to Deep Blue or AlphaGo is negligible—all one has to do is make the right moves.

---

* Omohundro, "The Basic AI Drives," in *Proceedings of the First AGI Conference*, 171; and in P. Wang, B. Goertzel, and S. Franklin, ed., *Artificial General Intelligence* (Amsterdam, The Netherlands: IOS Press, 2008).
† AI researcher Jeff Hawkins, for example, writes, "Some intelligent machines will be virtual, meaning they will exist and act solely within computer networks. . . . It is always possible to turn off a computer network, even if painful," https://www.recode.net/2015/3/2/11559576/.

- *Human-level or superhuman AI is impossible.** This is an unusual
  claim for AI researchers to make, given that, from Turing onward,
  they have been fending off such claims from philosophers and
  mathematicians. The claim, which is backed by no evidence, appears to concede that if superintelligent AI *were* possible, it *would*
  be a significant risk. It's as if a bus driver, with all of humanity as
  passengers, said, "Yes, I am driving toward a cliff—in fact, I'm
  pressing the pedal to the metal! But trust me, we'll run out of gas
  before we get there!" The claim represents a foolhardy bet against
  human ingenuity. We have made such bets before and lost. On September 11, 1933, renowned physicist Ernest Rutherford stated, with
  utter confidence, "Anyone who expects a source of power from the
  transformation of these atoms is talking moonshine." On September 12, 1933, Leo Szilard invented the neutron-induced nuclear
  chain reaction. A few years later he demonstrated such a reaction in
  his laboratory at Columbia University. As he recalled in a memoir:
  "We switched everything off and went home. That night, there was
  very little doubt in my mind that the world was headed for grief."
- *It's too soon to worry about it.* The right time to worry about a potentially serious problem for humanity depends not just on when the
  problem will occur but also on how much time is needed to devise
  and implement a solution that avoids the risk. For example, if we
  were to detect a large asteroid predicted to collide with the Earth in
  2067, would we say, "It's too soon to worry"? And if we consider the
  global catastrophic risks from climate change predicted to occur
  later in this century, is it too soon to take action to prevent them?
  On the contrary, it may be too late. The relevant timescale for
  human-level AI is less predictable, but, like nuclear fission, it might
  arrive considerably sooner than expected. One variation on this argument is Andrew Ng's statement that it's "like worrying about
  overpopulation on Mars." This appeals to a convenient analogy:

---

* The AI100 report (Peter Stone et al.), sponsored by Stanford University, includes the following: "Unlike in
the movies, there is no race of superhuman robots on the horizon or probably even possible," https://ai100
.stanford.edu/2016-report.

Not only is the risk easily managed and far in the future, but it also is extremely unlikely that we'd even try to move billions of humans to Mars in the first place. The analogy is a false one, however. We are already devoting huge scientific and technical resources to creating ever-more-capable AI systems. A more apt analogy would be a plan to move the human race to Mars with no consideration for what we might breathe, drink, or eat once we'd arrived.

- *Human-level AI isn't really imminent, in any case.* The AI100 report, for example, assures us, "Contrary to the more fantastic predictions for AI in the popular press, the Study Panel found no cause for concern that AI is an imminent threat to humankind." This argument simply misstates the reasons for concern, which are not predicated on imminence. In his 2014 book, *Superintelligence: Paths, Dangers, Strategies,* Nick Bostrom, for one, writes, "It is no part of the argument in this book that we are on the threshold of a big breakthrough in artificial intelligence, or that we can predict with any precision when such a development might occur."

- *You're just a Luddite.* It's an odd definition of Luddite that includes Turing, Wiener, Minsky, Musk, and Gates, who rank among the most prominent contributors to technological progress in the 20th and 21st centuries.* Furthermore, the epithet represents a complete misunderstanding of the nature of the concerns raised and the purpose for raising them. It is as if one were to accuse nuclear engineers of Luddism if they pointed out the need for control of the fission reaction. Some objectors also use the term "anti-AI," which is rather like calling nuclear engineers "anti-physics." The purpose of understanding and preventing the risks of AI is to ensure that we can realize the benefits. Bostrom, for example, writes that success in controlling AI will result in "a civilizational trajectory that leads to a compassionate and jubilant use of humanity's cosmic endowment"— hardly a pessimistic prediction.

---

* Elon Musk, Stephen Hawking, and others (including, apparently, the author) received the 2015 Luddite of the Year Award from the Information Technology Innovation Foundation, https://itif.org/publications/2016 /01/19/artificial-intelligence-alarmists-win-itif%E2%80%99s-annual-luddite-award.

- *Any machine intelligent enough to cause trouble will be intelligent enough to have appropriate and altruistic objectives.** (Often, the argument adds the premise that people of greater intelligence tend to have more altruistic objectives, a view that may be related to the self-conception of those making the argument.) This argument is related to Hume's is-ought problem and G. E. Moore's naturalistic fallacy, suggesting that somehow the machine, as a result of its intelligence, will simply *perceive* what is right, given its experience of the world. This is implausible; for example, one cannot perceive, in the design of a chessboard and chess pieces, the goal of checkmate; the same chessboard and pieces can be used for suicide chess, or indeed many other games still to be invented. Put another way: Where Bostrom imagines humans driven extinct by a putative robot that turns the planet into a sea of paper clips, we humans see this outcome as tragic, whereas the iron-eating bacterium *Thiobacillus ferrooxidans* is thrilled. Who's to say the bacterium is wrong? The fact that a machine has been given a fixed objective by humans doesn't mean that it will automatically recognize the importance to humans of things that aren't part of the objective. Maximizing the objective may well cause problems for humans, but, by definition, the machine will not recognize those problems as problematic.

- *Intelligence is multidimensional, "so 'smarter than humans' is a meaningless concept."†* It is a staple of modern psychology that IQ doesn't do justice to the full range of cognitive skills that humans possess to varying degrees. IQ is indeed a crude measure of human intelligence, but it is utterly meaningless for current AI systems, because their capabilities across different areas are uncorrelated. How do we compare the IQ of Google's search engine, which cannot play chess, with that of Deep Blue, which cannot answer search queries?

---

* Rodney Brooks, for example, asserts that it's impossible for a program to be "smart enough that it would be able to invent ways to subvert human society to achieve goals set for it by humans, without understanding the ways in which it was causing problems for those same humans," http://rodneybrooks.com/the -seven-deadly-sins-of-predicting-the-future-of-ai.
† Kevin Kelly, "The Myth of a Superhuman AI," *Wired*, April 25, 2017.

- None of this supports the argument that because intelligence is multifaceted, we can ignore the risk from superintelligent machines. If "smarter than humans" is a meaningless concept, then "smarter than gorillas" is also meaningless, and gorillas therefore have nothing to fear from humans; clearly, that argument doesn't hold water. Not only is it logically possible for one entity to be more capable than another across all the relevant dimensions of intelligence, it is also possible for one species to represent an existential threat to another even if the former lacks an appreciation for music and literature.

SOLUTIONS

Can we tackle Wiener's warning head-on? Can we design AI systems whose purposes don't conflict with ours, so that we're sure to be happy with how they behave? On the face of it, this seems hopeless, because it will doubtless prove infeasible to write down our purposes correctly or imagine all the counterintuitive ways a superintelligent entity might fulfill them.

If we treat superintelligent AI systems as if they were black boxes from outer space, then indeed we have no hope. Instead, the approach we seem obliged to take, if we are to have any confidence in the outcome, is to define some formal problem F, and design AI systems to be F solvers, such that no matter how perfectly a system solves F, we're guaranteed to be happy with the solution. If we can work out an appropriate F that has this property, we'll be able to create *provably beneficial* AI.

Here's an example of how *not* to do it: Let a reward be a scalar value provided periodically by a human to the machine, corresponding to how well the machine has behaved during each period, and let F be the problem of maximizing the expected sum of rewards obtained by the machine. The optimal solution to this problem is not, as one might hope, to behave well, but instead to take control of the human and force him or her to provide a stream of maximal rewards. This is known as the wireheading problem, based on observations that humans themselves are susceptible to the same problem if given a means to electronically stimulate their own pleasure centers.

There is, I believe, an approach that may work. Humans can reasonably be described as having (mostly implicit) preferences over their future lives—that is, given enough time and unlimited visual aids, a human could express a preference (or indifference) when offered a choice between two future lives laid out before him or her in all their aspects. (This idealization ignores the possibility that our minds are composed of subsystems with incompatible preferences; if true, that would limit a machine's ability to optimally satisfy our preferences, but it doesn't seem to prevent us from designing machines that avoid catastrophic outcomes.) The formal problem F to be solved by the machine in this case is to maximize human future-life preferences subject to its initial uncertainty as to what they are. Furthermore, although the future-life preferences are hidden variables, they're grounded in a voluminous source of evidence—namely, all of the human choices ever made. This formulation sidesteps Wiener's problem: The machine may learn more about human preferences as it goes along, of course, but it will never achieve complete certainty.

A more precise definition is given by the framework of cooperative inverse-reinforcement learning, or CIRL. A CIRL problem involves two agents, one human and the other a robot. Because there are two agents, the problem is what economists call a game. It is a game of partial information, because while the human knows the reward function, the robot doesn't—even though the robot's job is to maximize it.

A simple example: Suppose that Harriet, the human, likes to collect paper clips and staples and her reward function depends on how many of each she has. More precisely, if she has p paper clips and s staples, her degree of happiness is $\theta p + (1-\theta)s$, where θ *is* essentially an exchange rate between paper clips and staples. If θ is 1, she likes only paper clips; if θ is 0, she likes only staples; if θ is 0.5, she is indifferent between them; and so on. It's the job of Robby, the robot, to produce the paper clips and staples. The point of the game is that Robby wants to make Harriet happy, but he doesn't know the value of θ, so he isn't sure how many of each to produce.

Here's how the game works. Let the true value of θ be 0.49—that is, Harriet has a slight preference for staples over paper clips. And let's assume that Robby has a uniform prior belief about θ—that is, he believes θ is equally likely to be any value between 0 and 1. Harriet now gets to do a small demonstration, producing either two paper clips or two staples or one of each. After that, the robot can produce either ninety paper clips or ninety staples or fifty of each. You might think that Harriet, who prefers staples to paper clips, should produce two staples. But in that case, Robby's rational response would be to produce ninety staples (with a total value to Harriet of 45.9), which is a less desirable outcome for Harriet than fifty of each (total value 50.0). The optimal solution of this particular game is that Harriet produces one of each, so then Robby makes fifty of each. Thus, the way the game is defined encourages Harriet to "teach" Robby—as long as she knows that Robby is watching carefully.

Within the CIRL framework, one can formulate and solve the off-switch problem—that is, the problem of how to prevent a robot from disabling its off switch. (Turing may rest easier.) A robot that's uncertain about human preferences actually benefits from being switched off, because it understands that the human will press the off switch to prevent the robot from doing something counter to those preferences. Thus the robot is incentivized to preserve the off switch, and this incentive derives directly from its uncertainty about human preferences.*

The off-switch example suggests some templates for controllable-agent designs and provides at least one case of a provably beneficial system in the sense introduced above. The overall approach resembles mechanism-design problems in economics, wherein one incentivizes other agents to behave in ways beneficial to the designer. The key difference here is that we are building one of the agents in order to benefit the other.

There are reasons to think this approach may work in practice. First, there is abundant written and filmed information about humans doing

---

* See Hadfield-Menell et al., "The Off-Switch Game," https://arxiv.org/pdf/1611.08219.pdf.

things (and other humans reacting). Technology to build models of human preferences from this storehouse will presumably be available long before superintelligent AI systems are created. Second, there are strong, near-term economic incentives for robots to understand human preferences: If one poorly designed domestic robot cooks the cat for dinner, not realizing that its sentimental value outweighs its nutritional value, the domestic-robot industry will be out of business.

There are obvious difficulties, however, with an approach that expects a robot to learn underlying preferences from human behavior. Humans are irrational, inconsistent, weak willed, and computationally limited, so their actions don't always reflect their true preferences. (Consider, for example, two humans playing chess. Usually, one of them loses, but not on purpose!) So robots can learn from nonrational human behavior only with the aid of much better cognitive models of humans. Furthermore, practical and social constraints will prevent all preferences from being maximally satisfied simultaneously, which means that robots must mediate among conflicting preferences—something that philosophers and social scientists have struggled with for millennia. And what should robots learn from humans who enjoy the suffering of others? It may be best to zero out such preferences in the robots' calculations.

Finding a solution to the AI control problem is an important task; it may be, in Bostrom's words, "the essential task of our age." Up to now, AI research has focused on systems that are better at making decisions, but this is not the same as making better decisions. No matter how excellently an algorithm maximizes, and no matter how accurate its model of the world, a machine's decisions may be ineffably stupid in the eyes of an ordinary human if its utility function is not well aligned with human values.

This problem requires a change in the definition of AI itself—from a field concerned with pure intelligence, independent of the objective, to a field concerned with systems that are provably beneficial for humans. Taking the problem seriously seems likely to yield new ways of thinking about AI, its purpose, and our relationship to it.

Chapter 4

THE THIRD LAW

GEORGE DYSON

George Dyson is a historian of science and technology and the author of Baidarka: The Kayak, Darwin Among the Machines, Project Orion, *and* Turing's Cathedral.

In 2005, **George Dyson**, a historian of science and technology, visited Google at the invitation of some Google engineers. The occasion was the sixtieth anniversary of John von Neumann's proposal for a digital computer. After the visit, George wrote an essay, "Turing's Cathedral," which, for the first time, alerted the public about what Google's founders had in store for the world. "We are not scanning all those books to be read by people," explained one of his hosts after his talk. "We are scanning them to be read by an AI."

George offers a counternarrative to the digital age. His interests have included the development of the Aleut kayak, the evolution of digital computing and telecommunications, the origins of the digital universe, and a path not taken into space. His career (he never finished high school, yet has been awarded an honorary doctorate from the University of Victoria) has proved as impossible to classify as his books.

He likes to point out that analog computing, once believed to be as extinct as the Differential Analyzer, has returned. He argues that while we may use digital components, at a certain point the analog computing being performed by the system far exceeds the complexity of the digital code with which it is built. He believes that true artificial intelligence—with analog control systems emerging from a digital substrate the way digital computers emerged out of analog components in the aftermath of World War II—may not be as far off as we think.

In this essay, George contemplates the distinction between analog and digital computation and finds analog to be alive and well. Nature's response to an attempt to program machines to control everything may be machines without programming over which no one has control.

The history of computing can be divided into an Old Testament and a New Testament: before and after electronic digital computers and the codes they spawned proliferated across the Earth. The Old Testament prophets, who delivered the underlying logic, included Thomas Hobbes and Gottfried Wilhelm Leibniz. The New Testament prophets included Alan Turing, John von Neumann, Claude Shannon, and Norbert Wiener. They delivered the machines.

Alan Turing wondered what it would take for machines to become intelligent. John von Neumann wondered what it would take for machines to self-reproduce. Claude Shannon wondered what it would take for machines to communicate reliably, no matter how much noise intervened. Norbert Wiener wondered how long it would take for machines to assume control.

Wiener's warnings about control systems beyond human control appeared in 1949, just as the first generation of stored-program electronic digital computers were introduced. These systems required direct supervision by human programmers, undermining his concerns. What's the problem, as long as programmers are in control of the machines? Ever since, debate over the risks of autonomous control has remained associated with the debate over the powers and limitations of digitally coded machines. Despite their astonishing powers, little real autonomy has been observed. This is a dangerous assumption. What if digital computing is being superseded by something else?

Electronics underwent two fundamental transitions over the past hundred years: from analog to digital and from vacuum tubes to solid state. That these transitions occurred together does not mean they are inextricably linked. Just as digital computation was implemented using vacuum tube components, analog computation can be implemented in solid state. Analog computation is alive and well, even though vacuum tubes are commercially extinct.

There is no precise distinction between analog and digital computing. In general, digital computing deals with integers, binary sequences, deterministic logic, and time that is idealized into discrete increments, whereas analog computing deals with real numbers, nondeterministic logic, and continuous functions, including time as it exists as a continuum in the real world.

Imagine you need to find the middle of a road. You can measure its width using any available increment and then digitally compute the middle to the nearest increment. Or you can use a piece of string as an analog computer, mapping the width of the road to the length of the string and finding the middle, without being limited to increments, by doubling the string back upon itself.

Many systems operate across both analog and digital regimes. A tree integrates a wide range of inputs as continuous functions, but if you cut down that tree, you find that it has been counting the years digitally all along.

In analog computing, complexity resides in network topology, not in code. Information is processed as continuous functions of values such as voltage and relative pulse frequency rather than by logical operations on discrete strings of bits. Digital computing, intolerant of error or ambiguity, depends upon error correction at every step along the way. Analog computing tolerates errors, allowing you to live with them.

Nature uses digital coding for the storage, replication, and recombination of sequences of nucleotides, but relies on analog computing, running on nervous systems, for intelligence and control. The genetic system in every living cell is a stored-program computer. Brains aren't.

Digital computers execute transformations between two species of bits: bits representing differences in space and bits representing differences in time. The transformations between these two forms of information, sequence and structure, are governed by the computer's programming, and as long as computers require human programmers, we retain control.

Analog computers also mediate transformations between two forms of information: structure in space and behavior in time. There is no code and no programming. Somehow—and we don't fully understand how—nature evolved analog computers known as nervous systems, which embody information absorbed from the world. They learn. One of the things they learn is control. They learn to control their own behavior, and they learn to control their environment to the extent that they can.

Computer science has a long history—going back to before there even *was* computer science—of implementing neural networks, but for the most part these have been simulations of neural networks by digital computers, not neural networks as evolved in the wild by nature herself. This is starting to change: from the bottom up, as the threefold drivers of drone warfare, autonomous vehicles, and cell phones push the development of neuromorphic microprocessors that implement actual neural networks, rather than simulations of neural networks, directly in silicon (and other potential substrates); and from the top down, as our largest and most successful enterprises increasingly turn to analog computation in their infiltration and control of the world.

While we argue about the intelligence of digital computers, analog computing is quietly supervening upon the digital, in the same way that analog components like vacuum tubes were repurposed to build digital computers in the aftermath of World War II. Individually deterministic finite-state processors, running finite codes, are forming large-scale, nondeterministic, non-finite-state metazoan organisms running wild in the real world. The resulting hybrid analog/digital systems treat streams of bits collectively, the way the flow of electrons is treated in a vacuum tube, rather than individually, as bits are treated by the discrete-state devices generating the flow. Bits are the new electrons. Analog is back, and its nature is to assume control.

Governing everything from the flow of goods to the flow of traffic to the flow of ideas, these systems operate statistically, as pulse-frequency coded information is processed in a neuron or a brain. The emergence of intelligence gets the attention of *Homo sapiens,* but what we should be worried about is the emergence of control.

I magine it is 1958 and you are trying to defend the continental United States against airborne attack. To distinguish hostile aircraft, one of the things you need, besides a network of computers and early-warning radar sites, is a map of all commercial air traffic, updated in real time. The United States built such a system and named it SAGE (Semi-Automatic Ground Environment). SAGE in turn spawned Sabre, the first integrated reservation system for booking airline travel in real time. Sabre and its progeny soon became not just a map of what seats were available but also a system that began to control, with decentralized intelligence, where airliners would fly, and when.

But isn't there a control room somewhere, with someone at the controls? Maybe not. Say, for example, you build a system to map highway traffic in real time, simply by giving cars access to the map in exchange for reporting their own speed and location at the time. The result is a fully decentralized control system. Nowhere is there any controlling model of the system except the system itself.

Imagine it is the first decade of the 21st century and you want to track the complexity of human relationships in real time. For social life at a small college, you could construct a central database and keep it up to date, but its upkeep would become overwhelming if taken to any larger scale. Better to pass out free copies of a simple semi-autonomous code, hosted locally, and let the social network update itself. This code is executed by digital computers, but the analog computing performed by the system as a whole far exceeds the complexity of the underlying code. The resulting pulse-frequency coded model of the social graph *becomes* the social graph. It spreads wildly across the campus and then the world.

What if you wanted to build a machine to capture what everything known to the human species *means*? With Moore's Law behind you, it doesn't take too long to digitize all the information in the world. You scan every book ever printed, collect every email ever written, and gather forty-nine years of video every twenty-four hours, while tracking where people are and what they do, in real time. But how do you capture the *meaning*?

Even in the age of all things digital, this cannot be defined in any strictly logical sense, because meaning, among humans, isn't fundamentally logical. The best you can do, once you have collected all possible answers, is to invite well-defined questions and compile a pulse-frequency weighted map of how everything connects. Before you know it, your system will not only be observing and mapping the meaning of things, it will start *constructing* meaning as well. In time, it will *control* meaning, in the same way the traffic map starts to control the flow of traffic even though no one seems to be in control.

There are three laws of artificial intelligence. The first, known as Ashby's Law, after cybernetician W. Ross Ashby, author of *Design for a Brain*, states that any effective control system must be as complex as the system it controls.

The second law, articulated by John von Neumann, states that the defining characteristic of a complex system is that it constitutes its own simplest behavioral description. The simplest complete model of an organism is the organism itself. Trying to reduce the system's behavior to any formal description makes things more complicated, not less.

The third law states that any system simple enough to be understandable will not be complicated enough to behave intelligently, while any system complicated enough to behave intelligently will be too complicated to understand.

The third law offers comfort to those who believe that until we understand intelligence, we need not worry about superhuman intelligence arising among machines. But there is a loophole in the third law. It is entirely possible to build something without understanding it. You don't need to fully understand how a brain works in order to build one that works. This is a loophole that no amount of supervision over algorithms by programmers and their ethical advisers can ever close. Provably "good" AI is a myth. Our relationship with true AI will always be a matter of faith, not proof.

We worry too much about machine intelligence and not enough about self-reproduction, communication, and control. The next revolution in computing will be signaled by the rise of analog systems over which digital programming no longer has control. Nature's response to those who believe they can build machines to control everything will be to allow them to build a machine that controls them instead.

Chapter 5

WHAT CAN WE DO?

DANIEL C. DENNETT

Daniel C. Dennett is University Professor and Austin B. Fletcher Professor of Philosophy and co-director of the Center for Cognitive Studies at Tufts University. He is the author of a dozen books, including Consciousness Explained *and, most recently,* From Bacteria to Bach and Back: The Evolution of Minds.

Dan Dennett is the philosopher of choice in the AI community. He is perhaps best known in cognitive science for his concept of intentional systems and his model of human consciousness, which sketches a computational architecture for realizing the stream of consciousness in the massively parallel cerebral cortex. That uncompromising computationalism has been opposed by philosophers such as John Searle, David Chalmers, and the late Jerry Fodor, who have protested that the most important aspects of consciousness—intentionality and subjective qualia—cannot be computed.

Twenty-five years ago, I was visiting Marvin Minsky, one of the original AI pioneers, and asked him about Dan. "He's our best current philosopher—the next Bertrand Russell," said Marvin, adding that unlike traditional philosophers, Dan was a student of neuroscience, linguistics, artificial intelligence, computer science, and psychology.

"He's redefining and reforming the role of the philosopher. Of course, Dan doesn't understand my Society of Mind theory, but nobody's perfect."

Dan's view of the efforts of AI researchers to create superintelligent AIs is relentlessly levelheaded. What, me worry? In this essay, he reminds us that AIs, above all, should be regarded—and treated—as tools and not as humanoid colleagues.

He has been interested in information theory since his graduate school days at Oxford. In fact, he told me that early in his career he was keenly interested in writing a book about Wiener's cybernetic ideas. As a thinker who embraces the scientific method, one of his charms is his willingness to be wrong. Of a recent piece titled "What Is Information?" he has announced, "I stand by it, but it's under revision. I'm already moving beyond it and realizing there's a better way of tackling some of these issues." He will most likely remain cool and collected on the subject of AI research, although he has acknowledged, often, that his own ideas evolve—as anyone's ideas should.

Many have reflected on the irony of reading a great book when you are too young to appreciate it. Consigning a classic to the *already read* stack and thereby insulating yourself against any further influence while gleaning only a few ill-understood ideas from it is a recipe for neglect that is seldom benign. This struck me with particular force when I reread *The Human Use of Human Beings* more than sixty years after my juvenile encounter. We should all make it a regular practice to reread books from our youth, where we are apt to discover clear previews of some of our own later "discoveries" and "inventions," along with a wealth of insights to which we were bound to be impervious until our minds had been torn and tattered, exercised and enlarged, by confrontations with life's problems.

Writing at a time when vacuum tubes were still the primary electronic building blocks and there were only a few actual computers in operation, Norbert Wiener imagined the future we now contend with in impressive detail and with few clear mistakes. Alan Turing's famous 1950 article "Computing Machinery and Intelligence," in the philosophy journal *Mind,* foresaw the development of AI, and so did Wiener, but Wiener saw further and deeper, recognizing that AI would not just imitate—and replace—human beings in many intelligent activities but change human beings in the process:

> We are but whirlpools in a river of ever-flowing water. We are not
> stuff that abides, but patterns that perpetuate themselves.*

When that was written, it could be comfortably dismissed as yet another bit of Heraclitean overstatement. Yeah, yeah, you can never step in
the same river twice. But it contains the seeds of the revolution in outlook. Today we know how to think about complex adaptive systems,
strange attractors, extended minds, and homeostasis, a change in perspective that promises to erase the "explanatory gap"† between mind
and mechanism, spirit and matter, a gap that is still ardently defended by
latter-day Cartesians who cannot bear the thought that we—we
ourselves—are self-perpetuating patterns of information-bearing matter,
not "stuff that abides." Those patterns are remarkably resilient and self-
restoring but at the same time protean, opportunistic, selfish exploiters
of whatever new is available to harness in their quest for perpetuation.
And here is where things get dicey, as Wiener recognized. When attrac-
tive opportunities abound, we are apt to be willing to pay a little and
accept some small, even trivial, cost-of-doing-business for access to new
powers. And pretty soon we become so dependent on our new tools that
we lose the ability to thrive without them. Options become obligatory.

It's an old, old story, with many well-known chapters in evolutionary
history. Most mammals can synthesize their own vitamin C, but pri-
mates, having opted for a diet composed largely of fruit, lost the innate
ability. We are now obligate ingesters of vitamin C, but not obligate fru-
givores like our primate cousins, since we have opted for technology that
allows us to make, and take, vitamins as needed. The self-perpetuating
patterns that we call human beings are now dependent on clothes,
cooked food, vitamins, vaccinations, credit cards, smartphones, and the
Internet. And—tomorrow if not already today—AI.

Wiener foresaw the problems that Turing and the other optimists
have largely overlooked. The real danger, he said, is

---

* *The Human Use of Human Beings* (Boston: Houghton Mifflin, 1954), 96.
† Joseph Levine, "Materialism and Qualia: The Explanatory Gap," *Pacific Philosophical Quarterly* 64 (1983):
354–61.

that such machines, though helpless by themselves, may be used by a human being or a block of human beings to increase their control over the rest of the race or that political leaders may attempt to control their populations by means not of machines themselves but through political techniques as narrow and indifferent to human possibility as if they had, in fact, been conceived mechanically.*

The power, he recognized, lay primarily in the algorithms, not in the hardware they run on, although the hardware of today makes practically possible algorithms that would have seemed preposterously cumbersome in Wiener's day. What can we say about these "techniques" that are "narrow and indifferent to human possibility"? They have been introduced again and again, some obviously benign, some obviously dangerous, and many in the omnipresent middle ground of controversy.

Consider a few of the skirmishes. My late friend Joe Weizenbaum, Wiener's successor as MIT's Jeremiah of hi-tech, loved to observe that credit cards, whatever their virtues, also provided an inexpensive and almost foolproof way for the government, or corporations, to track the travels and habits and desires of individuals. The anonymity of cash has been largely underappreciated, except by drug dealers and other criminals, and now it may be going extinct. This may make money laundering a more difficult technical challenge in the future, but the AI pattern finders arrayed against it have the side effect of making us all more transparent to any "block of human beings" that may "attempt to control" us.

Looking to the arts, the innovation of digital audio and video recording lets us pay a small price (in the eyes of all but the most ardent audiophiles and film lovers) when we abandon analog formats, and in return provides easy—all too easy?—reproduction of artworks with almost perfect fidelity. But there is a huge hidden cost. Orwell's Ministry of Truth is now a practical possibility. AI techniques for creating all-but-undetectable forgeries of "recordings" of encounters are now becoming available, which will render obsolete the tools of investigation we have

* *The Human Use of Human Beings* (Boston: Houghton Mifflin, 1954), 181.

come to take for granted in the last hundred and fifty years. Will we simply abandon the brief Age of Photographic Evidence and return to the earlier world in which human memory and trust provided the gold standard, or will we develop new techniques of defense and offense in the arms race of truth? (We can imagine a return to *analog* film-exposed-to-light, kept in "tamper-proof" systems until shown to juries, etc., but how long would it be before somebody figured out a way to infect such systems with doubt? One of the disturbing lessons of recent experience is that the task of destroying a reputation for credibility is much less expensive than the task of protecting such a reputation.) Wiener saw the phenomenon at its most general: "[I]n the long run, there is no distinction between arming ourselves and arming our enemies." The information age is also the dysinformation age.

What can we do? We need to rethink our priorities with the help of the passionate but flawed analyses of Wiener, Weizenbaum, and the other serious critics of our technophilia. A key phrase, it seems to me, is Wiener's almost offhand observation, above, that "these machines" are "helpless by themselves." As I have been arguing recently, we're making tools, not colleagues, and the great danger is not appreciating the difference, which we should strive to accentuate, marking and defending it with political and legal innovations.

Perhaps the best way to see what is being missed is to note that Alan Turing himself suffered an entirely understandable failure of imagination in his formulation of the famous Turing Test. As everyone knows, it is an adaptation of his "imitation game," in which a man, hidden from view and communicating verbally with a judge, tries to convince the judge that he is in fact a woman, while a woman, also hidden and communicating with the judge, tries to convince the judge that she is the woman. Turing reasoned that this would be a demanding challenge for a man (or for a woman pretending to be a man), exploiting a wealth of knowledge about how the other sex thinks and acts, what they tend to favor or ignore. Surely (*ding!*)* any man who could beat a woman at being perceived to be

---

* The "surely" alarm (the habit of having a bell ring in your head whenever you see the word in an argument) is described and defended by me in *Intuition Pumps and Other Tools for Thinking* (New York: W. W. Norton, 2013).

a woman would be an intelligent agent. What Turing did not foresee is the power of deep-learning AI to acquire this wealth of information in an exploitable form *without having to understand it*. Turing imagined an astute and imaginative (and hence conscious) agent who cunningly designed his responses based on his detailed "theory" of what women are likely to do and say. Top-down intelligent design, in short. He certainly didn't think that a man, winning the imitation game, would somehow *become* a woman; he imagined that there would still be a man's consciousness guiding the show. The hidden premise in Turing's almost-argument was: Only a conscious, intelligent *agent* could devise and control a winning strategy in the imitation game. And so it was persuasive to Turing (and others, including me, still a stalwart defender of the Turing Test) to argue that a "computing machine" that could pass as human in a contest with a human might not be conscious in just the way a human being is, but would nevertheless have to be a conscious agent of *some* kind. I think this is still a defensible position—the only defensible position—but you have to understand how resourceful and ingenious a judge would have to be to expose the shallowness of the façade that a deep-learning AI (a tool, not a colleague) could present.

What Turing didn't foresee is the uncanny ability of superfast computers to sift mindlessly through Big Data, of which the Internet provides an inexhaustible supply, finding probabilistic patterns in human activity that could be used to pop "authentic"-seeming responses into the output for almost any probe a judge would think to offer. Wiener also underestimates this possibility, seeing the telltale weakness of a machine in not being able to

> take into account the vast range of probability that characterizes the human situation.*

But taking into account that range of probability is just where the new AI excels. The only chink in the armor of AI is that word "vast";

---

* *The Human Use of Human Beings* (Boston: Houghton Mifflin, 1954), 181.

human possibilities, thanks to language and the culture that it spawns, are truly Vast.* No matter how many patterns we may find with AI in the flood of data that has so far found its way onto the Internet, there are Vastly more possibilities that have never been recorded there. Only a fraction (but not a Vanishing fraction) of the world's accumulated wisdom and design and repartee and silliness has made it onto the Internet, but probably a better tactic for the judge to adopt when confronting a candidate in the Turing Test is not to *search* for such items but to *create* them anew. AI in its current manifestations is parasitic on human intelligence. It quite indiscriminately gorges on whatever has been produced by human creators and extracts the patterns to be found there—including some of our most pernicious habits.† These machines do not (yet) have the goals or strategies or capacities for self-criticism and innovation to permit them to transcend their databases by reflectively thinking about their own thinking and their own goals. They are, as Wiener says, helpless, not in the sense of being *shackled* agents or *disabled* agents but in the sense of not being agents at all—not having the capacity to be "moved by reasons" (as Kant put it) presented to them. It is important that we keep it that way, which will take some doing.

One of the flaws in Weizenbaum's book *Computer Power and Human Reason,* something I tried in vain to convince him of in many hours of discussion, is that he could never decide which of two theses he wanted to defend: *AI is impossible!* or *AI is possible but evil!* He wanted to argue, with John Searle and Roger Penrose, that "Strong AI" is impossible, but there are no good arguments for that conclusion. After all, everything we now know suggests that, as I have put it, we are robots made of robots made of robots . . . down to the motor proteins and their ilk, with no magical ingredients thrown in along the way. Weizenbaum's more important and defensible message was that we should not strive to create

---

* In *Darwin's Dangerous Idea* (New York: Simon & Schuster, 1995), p. 109, I coined the capitalized version, "Vast," meaning *Very much more than astronomical,* and its complement, "Vanishing," to replace the usual exaggerations *infinite* and *infinitesimal* for discussions of those possibilities that are not officially infinite but nevertheless infinite for all practical purposes.
† Aylin Caliskan-Islam, Joanna J. Bryson, and Arvind Narayanan, "Semantics Derived Automatically from Language Corpora Contain Human-Like Biases," *Science* 356, no. 6334 (April 14, 2017): 183–86, DOI: 10.1126/science.aal4230.

Strong AI and should be extremely cautious about the AI systems that we can create and have already created. As one might expect, the defensible thesis is a hybrid: *AI (Strong AI) is possible in principle but not desirable. The AI that's practically possible is not necessarily evil—unless it is mistaken for Strong AI!*

The gap between today's systems and the science-fictional systems dominating the popular imagination is still huge, though many folks, both lay and expert, manage to underestimate it. Let's consider IBM's Watson, which can stand as a worthy landmark for our imaginations for the time being. It is the result of a very large-scale R&D process extending over many person-centuries of intelligent design, and as George Church notes in these pages, it uses thousands of times more energy than a human brain (a technological limitation that, as he also notes, may be temporary). Its victory in *Jeopardy!* was a genuine triumph, made possible by the formulaic restrictions of the *Jeopardy!* rules, but in order for it to compete, even these rules had to be revised (one of those trade-offs: you give up a little versatility, a little humanity, and get a crowd-pleasing show). Watson is not good company, in spite of misleading ads from IBM that suggest a general conversational ability, and turning Watson into a plausibly multidimensional *agent* would be like turning a hand calculator into Watson. Watson could be a useful core faculty for such an agent, but more like a cerebellum or an amygdala than a mind—at best, a special-purpose subsystem that could play a big supporting role, but not remotely up to the task of framing purposes and plans and building insightfully on its conversational experiences.

Why would we want to create a thinking, creative agent out of Watson? Perhaps Turing's brilliant idea of an operational test has lured us into a trap: the quest to create at least the illusion of a real person behind the screen, bridging the "uncanny valley." The danger here is that ever since Turing posed his challenge—which was, after all, a challenge to *fool* the judges—AI creators have attempted to paper over the valley with cutesy humanoid touches, Disneyfication effects that will enchant and disarm the uninitiated. Weizenbaum's ELIZA was the pioneer example of such superficial illusion making, and it was his dismay at the ease with which his laughably simple and shallow program could persuade people they were having a serious heart-to-heart conversation that first sent him on his mission.

He was right to be worried. If there is one thing we have learned from the restricted Turing Test competitions for the Loebner Prize, it is that even very intelligent people who aren't tuned in to the possibilities and shortcuts of computer programming are readily taken in by simple tricks. The attitudes of people in AI toward these methods of dissembling at the "user interface" have ranged from contempt to celebration, with a general appreciation that the tricks are not deep but can be potent. One shift in attitude that would be very welcome is a candid acknowledgment that humanoid embellishments are *false advertising*—something to condemn, not applaud.

How could that be accomplished? Once we recognize that people are starting to make life-or-death decisions largely on the basis of "advice" from AI systems whose inner operations are unfathomable in practice, we can see a good reason why those who in any way encourage people to put more trust in these systems than they warrant should be held morally and legally accountable. AI systems are very powerful tools—so powerful that even experts will have good reason not to trust their own judgment over the "judgments" delivered by their tools. But then, if these tool users are going to benefit, financially or otherwise, from driving these tools through *terra incognita,* they need to make sure they know how to do this responsibly, with maximum control and justification. Licensing and bonding operators, just as we license pharmacists (and crane operators!) and other specialists whose errors and misjudgments can have dire consequences, can, with pressure from insurance companies and other underwriters, oblige creators of AI systems to go to extraordinary lengths to search for and reveal weaknesses and gaps in their products, and to train those entitled to operate them.

One can imagine a sort of inverted Turing Test in which the judge is on trial; until he or she can spot the weaknesses, the overstepped boundaries, the gaps in a system, no license to operate will be issued. The mental training required to achieve certification as a judge will be demanding. The urge to adopt the intentional stance, our normal tactic whenever we encounter what seems to be an intelligent agent, is almost overpoweringly strong. Indeed, the capacity to resist the allure of treating an apparent person as a person is an ugly talent, reeking of racism or species-ism. Many people would find the cultivation of such a ruthlessly skeptical approach morally repugnant, and we can anticipate that even the most proficient system users would occasionally succumb to the temptation to "befriend" their tools, if only to assuage their discomfort with the execution of their duties. No matter how scrupulously the AI designers launder the phony "human" touches out of their wares, we can expect novel habits of thought, conversational gambits and ruses, traps and bluffs, to arise in this novel setting for human action. The comically long lists of known side effects of new drugs advertised on television will be dwarfed by the obligatory revelations of the sorts of questions that can*not* be responsibly answered by particular systems, with heavy penalties for those who "overlook" flaws in their products. It is widely noted that a considerable part of the growing economic inequality in today's world is due to the wealth accumulated by digital entrepreneurs; we should enact legislation that puts their deep pockets in escrow for the public good. Some of the deepest pockets are voluntarily out in front of these obligations to serve society first and make money secondarily, but we shouldn't rely on goodwill alone.

We don't need artificial conscious agents. There is a surfeit of natural conscious agents, enough to handle whatever tasks should be reserved for such special and privileged entities. We need intelligent tools. Tools do not have rights, and should not have feelings that could be hurt, or be able to respond with resentment to "abuses" rained on them by inept users.* One of the reasons for not making artificial conscious agents is that however autonomous they might become (and in principle, they can be as autonomous, as self-enhancing or self-creating, as any person), they

---

* Joanna J. Bryson, "Robots Should Be Slaves," In *Close Engagement with Artificial Companions*, Yorick Wilks, ed. (Amsterdam, The Netherlands: John Benjamins, 2010), 63–74; http://www.cs.bath.ac.uk/~jjb/ftp/Bryson -Slaves-Book09.html; Joanna J. Bryson, "Patiency Is Not a Virtue: AI and the Design of Ethical Systems," https://www.cs.bath.ac.uk/~jjb/ftp/Bryson-Patiency-AAAISS16.pdf.

would not—without special provision, which might be waived—share with us natural conscious agents our vulnerability or our mortality.

I once posed a challenge to students in a seminar at Tufts I co-taught with Matthias Scheutz on artificial agents and autonomy: Give me the specs for a robot that could sign a binding contract with you—not as a surrogate for some human owner but on its own. This isn't a question of getting it to understand the clauses or manipulate a pen on a piece of paper but of having and *deserving* legal status as a morally responsible agent. Small children can't sign such contracts, nor can those disabled people whose legal status requires them to be under the care and responsibility of guardians of one sort or another. The problem for robots who might want to attain such an exalted status is that, like Superman, they are too invulnerable to be able to make a credible promise. If they were to renege, what would happen? What would be the penalty for promise breaking? Being locked in a cell or, more plausibly, dismantled? Being locked up is barely an inconvenience for an AI unless we first install artificial wanderlust that cannot be ignored or disabled by the AI on its own (and it would be systematically difficult to make this a foolproof solution, given the presumed cunning and self-knowledge of the AI); and dismantling an AI (either a robot or a bedridden agent like Watson) is not killing it if the information stored in its design and software is preserved. The very ease of digital recording and transmitting— the breakthrough that permits software and data to be, in effect, immortal—removes robots from the world of the vulnerable (at least robots of the usually imagined sorts, with digital software and memories). If this isn't obvious, think about how human morality would be affected if we could make "backups" of people every week, say. Diving headfirst on Saturday off a high bridge without benefit of a bungee cord would be a rush that you wouldn't remember when your Friday night backup was put online Sunday morning, but you could enjoy the videotape of your apparent demise thereafter.

So what we are creating are not—should not be—conscious, humanoid agents but an entirely new sort of entity, rather like oracles, with no conscience, no fear of death, no distracting loves and hates, no personality (but all sorts of foibles and quirks that would no doubt be identified as the "personality" of the system): boxes of truths (if we're lucky) almost certainly contaminated with a scattering of falsehoods. It will be hard enough learning to live with them without distracting ourselves with fantasies about the Singularity in which these AIs will enslave us, literally. The *human* use of human beings will soon be changed—once again—forever, but we can take the tiller and steer between some of the hazards if we take responsibility for our trajectory.

Chapter 6

THE INHUMAN MESS OUR MACHINES HAVE GOTTEN US INTO

RODNEY BROOKS

Rodney Brooks is a computer scientist; Panasonic Professor of Robotics, emeritus, MIT; former director of the MIT Artificial Intelligence Laboratory and the MIT Computer Science & Artificial Intelligence Laboratory (CSAIL). He is the author of Flesh and Machines.

The roboticist **Rodney Brooks**, featured in Errol Morris's 1997 documentary *Fast, Cheap & Out of Control* along with a lion tamer, a topiarist, and an expert on the naked mole rat, was described by one reviewer as "smiling with a wild gleam in his eye." But that's pretty much true of most visionaries.

A few years later in his career, Brooks, as befits one of the world's leading roboticists, suggested that "we overanthropomorphize humans, who are after all mere machines." He went on to present a warmhearted vision of a coming AI world in which "the distinction between us and robots is going to disappear." He also admitted to something of a divided worldview. "Like a religious scientist, I maintain two sets of inconsistent beliefs and act on each of them in different circumstances," he wrote. "It is this transcendence between belief systems that I think will be what enables mankind to ultimately accept robots as emotional machines, and thereafter start to empathize with them and attribute free will, respect, and ultimately rights to them."

That was in 2002. In these pages, he takes a somewhat more jaundiced, albeit narrower, view; he is alarmed by the extent to which we have come to rely on pervasive systems that are not just exploitative but also vulnerable, as a result of the too-rapid development of software engineering—an advance that seems to have outstripped the imposition of reliably effective safeguards.

Mathematicians and scientists are often limited in how they see the big picture, beyond their particular field, by the tools and metaphors they use in their work. Norbert Wiener is no exception, and I might guess that neither am I.

When he wrote *The Human Use of Human Beings,* Wiener was straddling the end of the era of understanding machines and animals simply as physical processes and the beginning of our current era of understanding machines and animals as computational processes. I suspect there will be future eras whose tools will look as distinct from the tools of the two eras Wiener straddled as those tools did from each other.

Wiener was a giant of the earlier era and built on the tools developed since the time of Newton and Leibniz to describe and analyze continuous processes in the physical world. In 1948 he published *Cybernetics,* a word he coined to describe the science of communication and control in both machines and animals. Today we would refer to the ideas in this book as control theory, an indispensable discipline for the design and analysis of physical machines, while mostly neglecting Wiener's claims about the science of communication. Wiener's innovations were largely driven by his work during the Second World War on mechanisms to aim and fire antiaircraft guns. He brought mathematical rigor to the design of the sorts of technology whose design processes had been largely heuristic in nature: from the Roman waterworks through Watt's steam engine to the early development of automobiles.

One can imagine a different contingent version of our intellectual and technological history had Alan Turing and John von Neumann, both of whom made major contributions to the foundations of computing, not appeared on the scene. Turing contributed a fundamental model of computation—now known as a Turing Machine—in his paper "On Computable Numbers with an Application to the Entscheidungsproblem," written and revised in 1936 and published in 1937. In these machines, a linear tape of symbols from a finite alphabet encodes the input for a computational problem and also provides the working space for the computation. A different machine was required for each separate computational problem; later work by others would show that in one particular machine, now known as a Universal Turing Machine, an arbitrary set of computing instructions could be encoded on that same tape.

In the 1940s, von Neumann developed an abstract self-reproducing machine called a cellular automaton. In this case it occupied a finite subset of an infinite two-dimensional array of squares each containing a single symbol from a finite alphabet of twenty-nine distinct symbols—the rest of the infinite array starts out blank. The single symbols in each square change in lockstep, based on a complex but finite rule about the current symbol in that square and its immediate neighbors. Under the complex rule that von Neumann developed, most of the symbols in most of the squares stay the same and a few change at each step. So when one looks at the nonblank squares, it appears that there is a constant structure with some activity going on inside it. When von Neumann's abstract machine reproduced, it made a copy of itself in another region of the plane. Within the "machine" was a horizontal line of squares that acted as a finite linear tape, using a subset of the finite alphabet. It was the symbols in those squares that encoded the machine of which they were a part. During the machine's reproduction, the "tape" could move either left or right and was both interpreted (transcribed) as the instructions (translation) for the new "machine" being built and then copied (replicated)—with the new copy being placed inside the new machine for further reproduction. Francis Crick and James Watson later showed, in 1953, how such a tape could be instantiated in biology by a long DNA molecule with its finite alphabet of four nucleobases: guanine, cytosine, adenine, and thymine (G, C, A, and T).* As in von Neumann's machine, in biological reproduction the linear sequence of symbols in DNA is interpreted—through transcription into RNA molecules, which are then translated into proteins, the structures that make up a new cell—and the DNA is replicated and encased in the new cell.

A second foundational piece of work was in a 1945 "First Draft" report on the design for a digital computer, wherein von Neumann advocated for a memory that could contain both instructions and data.† This is now known as a von Neumann architecture computer—as distinct from a Harvard architecture computer, where there are two separate memories, one for instructions and one for data. The vast majority of computer chips built in the era of Moore's Law are based on the von Neumann architecture, including those powering our data centers, our laptops, and our smartphones. Von Neumann's digital-computer architecture is conceptually the same generalization—from early digital computers constructed with electromagnetic relays at both Harvard University and Bletchley Park—that occurs in going from a special-purpose Turing Machine to a Universal Turing Machine. Furthermore, his self-replicating automata share a fundamental similarity with both the construction of a Turing Machine and the mechanism of DNA-based reproducing biological cells. There is to this day scholarly debate over whether von Neumann saw the cross connections between these three pieces of work, Turing's and his two. Turing's revision of his paper was done while he and von Neumann were both at Princeton; indeed, after getting his PhD, Turing almost stayed on as von Neumann's postdoc.

Without Turing and von Neumann, the cybernetics of Wiener might have remained a dominant mode of thought and driver of technology for

---

* "A Structure for Deoxyribose Nucleic Acid," *Nature* 171 (1953): 737–38.
† J. von Neumann, "First Draft of a Report on the EDVAC," *IEEE Annals of the History of Computing* 15 (1993): 27–75. Von Neumann is listed as the only author, whereas others contributed to the concepts he laid out; thus credit for the architecture has gone to him alone.

much longer than its brief moment of supremacy. In this imaginary version of history, we might well live today in an actual steam-punk world and not just get to observe its fantastical instantiations at Maker Faires!

My point is that Wiener thought about the world—physical, biological, and (in *Human Use*) sociological—in a particular way. He analyzed the world as continuous variables, as he explains in chapter 1 along with a nod to thermodynamics through an overlay of Gibbs statistics. He also shoehorns in a weak and unconvincing model of information as message passing between and among both physical and biological entities. To me, and from today's vantage point seventy years on, his tools seem woefully inadequate for describing the mechanisms underlying biological systems, and so he missed out on how similar mechanisms might eventually be embodied in technological computational systems—as now they have been. Today's dominant technologies were developed in the world of Turing and von Neumann, rather than the world of Wiener.

In the first industrial revolution, energy from a steam engine or a waterwheel was used by human workers to replace their own energy. Instead of being a source of energy for physical work, people became modulators of how a large source of energy was used. But because steam engines and waterwheels had to be large to be an efficient use of capital, and because in the 18th century the only technology for spatial distribution of energy was mechanical and worked only at very short range, many workers needed to be crowded around the source of energy. Wiener correctly argues that the ability to transmit energy as electricity caused a second industrial revolution. Now the source of energy could be distant from where it was used, and from the beginning of the 20th century, manufacturing could be much more dispersed as electrical-distribution grids were built.

Wiener then argues that a further new technology, that of the nascent computational machines of his time, will provide yet another revolution. The machines he talks about seem to be both analog and (perhaps) digital in nature; and he points out, in *The Human Use of Human Beings*, that since they will be able to make decisions, both blue-collar and white-collar workers may be reduced to being mere cogs in a much bigger machine. He fears that humans might use and abuse one another through organizational structures that this capability will encourage. We have certainly seen this play out in the last sixty years, and that disruption is far from over.

However, his physics-based view of computation protected him from realizing just how bad things might get. He saw machines' ability to communicate as providing a new and more inhuman way of exerting command and control. He missed that within a few decades computation systems would become more like biological systems, and it seems, from his descriptions in chapter 10 of his own work on modeling some aspects of biology, that he woefully underappreciated the many orders of magnitude of further complexity of biology over physics. We are in a much more complex situation today than he foresaw, and I am worried that it is much more pernicious than even his worst imagined fears.

In the 1960s, computation became firmly based on the foundations set out by Turing and von Neumann, and it was digital computation, based on the idea of finite alphabets, which they both used. An arbitrarily long sequence, or string, formed by characters from a finite alphabet, can be encoded as a unique integer. As with Turing Machines, the formalism for computation became that of computing an integer-valued function of a single integer-valued input.

Turing and von Neumann both died in the 1950s, and at that time this is how they saw computation. Neither foresaw the exponential increase in computing capability that Moore's Law would bring—nor how pervasive computing machinery would become. Nor did they foresee two developments in our modeling of computation, each of which poses a great threat to human society.

The first is rooted in the abstractions they adopted. In the fifty-year Moore's Law–fueled race to produce software that could exploit the doubling of computer capability every two years, the typical care and certification of engineering disciplines were thrown by the wayside. Software engineering was fast and prone to failures. This rapid development of software without standards of correctness has opened up many routes to exploit von Neumann architecture's storage of data and instructions in the same memory. One of the most common routes, known as "buffer overrun," involves an input number (or long string of characters) that is bigger than the programmer expected and overflows into where the instructions are stored. By carefully designing an input number that is too big by far, someone using a piece of software can infect it with instructions not intended by the programmer, and thus change what it does. This is the basis for creating a computer virus—so named for its similarity to a biological virus. The latter injects extra DNA into a cell, and that cell's transcription and translation mechanism blindly interprets it, making proteins that may be harmful to the host cell. Furthermore, the replication mechanism for the cell takes care of multiplying the virus. Thus a small foreign entity can take control of a much bigger entity and bend its behavior in unexpected ways.

These and other forms of digital attacks have taken the security of our everyday lives from us. We rely on computers for almost everything now. We rely on computers for our infrastructure of electricity, gas, roads, cars, trains, and airplanes; these are all vulnerable. We rely on computers for our banking, our payment of bills, our retirement accounts, our mortgages, our purchasing of goods and services—these, too, are all vulnerable. We rely on computers for our entertainment, our communications both business and personal, our physical security at home, our information about the world, and our voting systems—all vulnerable. None of this will get fixed anytime soon. In the meantime, many aspects of our society are open to vicious attacks, whether by freelancing criminals or nation-state adversaries.

The second development is that computation has gone beyond simply computing functions. Instead, programs remain online continuously, and so they can gather data about a sequence of queries. Under the Wiener / Turing / von Neumann scheme, we might think of the communication pattern for a Web browser to be:

User: Give me Web page A.

Browser: Here is Web page A.

. . .

User: Give me Web page B.

Browser: Here is Web page B.

Now instead it can look like this:

User: Give me Web page A.

Browser: Here is Web page A. [And I will secretly remember that you asked for Web page A.]

. . .

User: Give me Web page B.

Browser: Here is Web page B. [I see a correlation between its contents and that of the earlier requested Web page A, so I will update my model of you, the user, and transmit it to the company that produced me.]

When the machine no longer simply computes a function but instead maintains a state, it can start to make inferences about the human by the sequence of requests presented to it. And when different programs correlate across different request streams—say, correlating Web-page searches with social-media posts, or the payment for services on another platform, or the dwell time on a particular advertisement, or where the user has walked or driven with their GPS-enabled smartphone—the total systems of many programs communicating with one another and with databases leads to a whole new loss of privacy. The great exploitative leap made by so many West Coast companies has been to monetize those inferences without the knowing permission of the person generating the interactions with the computing machine platforms.

Wiener, Turing, and von Neumann could not foresee the complexity of those platforms, wherein the legal mumbo jumbo of the terms-of-use contracts the humans willingly enter into, without an inkling of what they entail, leads them to give up rights they would never concede in a one-on-one interaction with another human being. The computation platforms have become a shield behind which some companies hide in order to inhumanly exploit others. In certain other countries, the governments carry out these manipulations, and there the goal is not profits but the suppression of dissent.

Humankind has gotten itself into a fine pickle: We are being exploited by companies that paradoxically deliver services we crave, and at the same time our lives depend on many software-enabled systems that are open to attack. Getting ourselves out of this mess will be a long-term project. It will involve engineering, legislation, and, most important, moral leadership. Moral leadership is the first and biggest challenge.

Chapter 7

THE UNITY OF INTELLIGENCE

FRANK WILCZEK

Frank Wilczek is Herman Feshbach Professor of Physics at MIT, recipient of the 2004 Nobel Prize in physics, and the author of A Beautiful Question: Finding Nature's Deep Design.

I first met **Frank Wilczek** in the 1980s, when he invited me to his home in Princeton to talk about anyons. "The address is 112 Mercer Street," he wrote. "Look for the house with no driveway." So there I was, a few hours later, in Einstein's old living room, talking to a future recipient of the Nobel Prize in physics. If Frank was as impressed as I was by the surroundings, you'd never guess it. His only comment concerned the difficulty of finding a parking place in front of a "house with no driveway."

Unlike most theoretical physicists, Frank has long had a keen interest in AI, as witnessed in these three "Observations":

1. "Francis Crick called it 'the Astonishing Hypothesis': that consciousness, also known as Mind, is an emergent property of matter," which, if true, indicates that "all intelligence is machine intelligence. What distinguishes natural from artificial intelligence is not what it is, but only how it is made."

2. "Artificial intelligence is not the product of an alien invasion.
   It is an artifact of a particular human culture and reflects the
   values of that culture."

3. "David Hume's striking statement 'Reason Is, and Ought only
   to Be, the Slave of the Passions' was written in 1738 [and] was, of
   course, meant to apply to human reason and human passions. . . .
   But Hume's logical/philosophical point remains valid for AI. Simply
   put: Incentives, not abstract logic, drive behavior."

He notes that "the big story of the 20th and the 21st century is
that [as] computing develops, we learn how to calculate the
consequences of the [fundamental] laws better and better. There's
also a feedback cycle: When you understand matter better, you can
design better computers, which will enable you to calculate better. It's
kind of an ascending helix."

Here he argues that human intelligence, for now, holds the
advantage—yet our future, unbounded by our solar system and
doubtless also by our galaxy, will never be realized without the help of
our AIs.

I. A SIMPLE ANSWER TO CONTENTIOUS QUESTIONS

- Can an artificial intelligence be conscious?
- Can an artificial intelligence be creative?
- Can an artificial intelligence be evil?

Those questions are often posed today, both in popular media and in scientifically informed debates. But the discussions never seem to converge. Here I'll begin by answering them as follows:

Based on physiological psychology, neurobiology, and physics, it would be very surprising if the answers were not Yes, Yes, and Yes. The reason is simple, yet profound: Evidence from those fields makes it overwhelmingly likely that there is no sharp divide between natural and artificial intelligence.

In his 1994 book of that title, the renowned biologist Francis Crick proposed an "Astonishing Hypothesis": that mind emerges from matter. He famously claimed that mind, in all its aspects, is "no more than the behavior of a vast assembly of nerve cells and their associated molecules."

The "Astonishing Hypothesis" is in fact the foundation of modern neuroscience. People try to understand how minds work by understanding how brains function; and they try to understand how brains function by studying how information is encoded in electrical and chemical signals, transformed by physical processes, and used to control behavior. In that scientific endeavor, they make no allowance for extraphysical behavior. So far, in thousands of exquisite experiments, that strategy has never failed. It has never proved necessary to allow for the influence of consciousness or creativity unmoored from brain activity to explain any observed fact of psychophysics or neurobiology. No one has ever stumbled upon a power of mind that is separate from conventional physical events in biological organisms. While there are many things we do not understand about brains, and about minds, the "astonishing hypothesis" has held intact.

If we broaden our view beyond neurobiology to consider the whole range of scientific experimentation, the case becomes still more compelling. In modern physics, the foci of interest are often extremely delicate phenomena. To investigate them, experimenters must take many precautions against contamination by "noise." They often find it necessary to construct elaborate shielding against stray electric and magnetic fields; to compensate for tiny vibrations due to microearthquakes or passing cars; to work at extremely low temperatures and in high vacuum; and so forth. But there's a notable exception: They have never found it necessary to make allowances for what people nearby (or, for that matter, far away) are thinking. No "thought waves," separate from known physical processes yet capable of influencing physical events, seem to exist.

That conclusion, taken at face value, erases the distinction between natural and artificial intelligence. It implies that if we were to duplicate, or accurately simulate, the physical processes occurring in a brain—as, in principle, we can—and wire up its input and output to sense organs and muscles, then we would reproduce, in a physical artifact, the observed manifestations of natural intelligence. Nothing observable would be missing. As an observer, I'd have no less (and no more) reason to ascribe consciousness, creativity, or evil to that artifact than I do to ascribe those properties to its natural counterparts, like other human beings.

Thus, by combining Crick's "astonishing hypothesis" in neurobiology with powerful evidence from physics, we deduce that natural intelligence is a special case of artificial intelligence. That conclusion deserves a name, and I will call it the "astonishing corollary."

With that, we have the answer to our three questions. Since consciousness, creativity, and evil are obvious features of natural human intelligence, they are possible features of artificial intelligence.

A hundred years ago, or even fifty, to believe the hypothesis that mind emerges from matter, and to infer our corollary that natural intelligence is a special case of artificial intelligence, would have been leaps of faith. In view of the many surrounding gaps—chasms, really—in contemporary understanding of biology and physics, they were genuinely doubtful propositions. But epochal developments in those areas have changed the picture:

In biology: A century ago, not only thought but also metabolism, heredity, and perception were deeply mysterious aspects of life that defied physical explanation. Today, of course, we have extremely rich and detailed accounts of metabolism, heredity, and many aspects of perception, from the bottom up, starting at the molecular level.

In physics: After a century of quantum physics and its application to materials, physicists have discovered, over and over, how rich and strange the behavior of matter can be. Superconductors, lasers, and many other wonders demonstrate that large assemblies of molecular units, each simple in itself, can exhibit qualitatively new, "emergent" behavior, while remaining fully obedient to the laws of physics. Chemistry, including biochemistry, is a cornucopia of emergent phenomena, all now quite firmly grounded in physics. The pioneering physicist Philip Anderson, in an essay titled "More Is Different," offers a classic discussion of emergence. He begins by acknowledging that "the reductionist hypothesis [i.e., the completeness of physical explanations based on known interactions of simple parts] may still be a topic for controversy among philosophers, but among the great majority of active scientists I think it is accepted without question." But he goes on to emphasize that "[t]he behavior of large and complex aggregates of elementary particles, it turns out, is not to be understood in terms of a simple extrapolation of the properties of a few particles."* Each new level of size and complexity supports new forms of organization, whose patterns encode information in new ways and whose behavior is best described using new concepts.

Electronic computers are a magnificent example of emergence. Here, all the cards are on the table. Engineers routinely design, from the bottom up, based on known (and quite sophisticated) physical principles, machines that process information in extremely impressive ways. Your iPhone can beat you at chess, quickly collect and deliver information about anything, and take great pictures, too. Because the process whereby computers, smartphones, and other intelligent objects are designed and manufactured is completely transparent, there can be no doubt that their wonderful capabilities emerge from regular physical processes, which we can trace down to the level of electrons, photons, quarks, and gluons. Evidently, brute matter can get pretty smart.

Let me summarize the argument. From two strongly supported hypotheses, we've drawn a straightforward conclusion:

- Human mind emerges from matter.
- Matter is what physics says it is.
- *Therefore,* the human mind emerges from physical processes we understand and can reproduce artificially.
- *Therefore,* natural intelligence is a special case of artificial intelligence.

Of course, our "astonishing corollary" could fail; the first two lines of this argument are hypotheses. But their failure would have to bring in a foundation-shattering discovery—a significant new phenomenon, with large-scale physical consequences, that takes place in unremarkable, well-studied physical circumstances (i.e., the materials, temperatures, and pressures inside human brains), yet that has somehow managed for

---

* *Science* 177, no. 4047 (August 4, 1972): 393–96.

many decades to elude determined investigators armed with sophisti-cated instruments. Such a discovery would be . . . astonishing.

II. THE FUTURE OF INTELLIGENCE

It is part of human nature to improve on human bodies and minds. His-torically, clothing, eyeglasses, and watches are examples of increasingly sophisticated augmentations that enhance our toughness, perception, and awareness. They are major improvements to the natural human en-dowment, whose familiarity should not blind us to their depth. Today smartphones and the Internet are bringing the human drive toward aug-mentation into realms more central to our identity as intelligent beings. They are giving us, in effect, quick access to a vast collective awareness and a vast collective memory.

At the same time, autonomous artificial intelligences have become world champions in a wide variety of "cerebral" games, such as chess and Go, and have taken over many sophisticated pattern-recognition tasks, such as reconstructing what happened during complex reactions at the Large Hadron Collider from a blizzard of emerging particle tracks to find new particles; or gathering clues from fuzzy X-ray, fMRI, and other types of images to diagnose medical problems.

Where is this drive toward self-enhancement and innovation taking us? While the precise sequence of events and the timescale over which they'll play out are impossible to predict (or, at least, beyond me), some basic considerations suggest that eventually the most powerful embodi-ments of mind will be quite different things from human brains as we know them today.

Consider six factors whereby information-processing technology ex-ceeds human capabilities—vastly, qualitatively, or both:

- *Speed:* The orchestrated motion of electrons, which is the heart of modern artificial information processing, can be much faster than the processes of diffusion and chemical change by which brains operate. Typical modern computer clock rates approach 10 gigahertz, corresponding to 10 billion operations per second. No single measure of speed applies to the bewildering variety of brain processes, but one fundamental limitation is latency of action potentials, which limits their spacing to a few 10s per second. It is probably no accident that the "frame rate," at which we can distinguish that movies are actually a sequence of stills, is about 40 per second. Thus, electronic processing is close to a billion times faster.

- *Size:* The linear dimension of a typical neuron is about 10 microns. Molecular dimensions, which set a practical limit, are about 10,000 times smaller, and artificial processing units are approaching that scale. Smallness makes communication more efficient.

- *Stability:* Whereas human memory is essentially continuous (analog), artificial memory can incorporate discrete (digital) features. Whereas analog quantities can erode, digital quantities can be stored, refreshed, and maintained with complete accuracy.

- *Duty cycle:* Human brains grow tired with effort. They need time off to take nourishment and to sleep. They carry the burden of aging. Most profoundly: They die.

- *Modularity* (open architecture): Because artificial information processors can support precisely defined digital interfaces, they can readily assimilate new modules. Thus, if we want a computer to "see" ultraviolet or infrared or "hear" ultrasound, we can feed the output from an appropriate sensor directly into its "nervous system." The architecture of brains is much more closed and opaque, and the human immune system actively resists implants.

- *Quantum readiness:* One case of modularity deserves special mention because of its long-term potential. Recently physicists and information scientists have come to appreciate that the principles of quantum mechanics support new computing principles, which can empower qualitatively new forms of information processing and (plausibly) new levels of intelligence. But these possibilities rely on aspects of quantum behavior that are quite delicate and seem especially unsuitable for interfacing with the warm, wet, messy environment of human brains.

Evidently, as platforms for intelligence, human brains are far from optimal. Still, although versatile housekeeping robots or mechanical soldiers would find ready, lucrative markets, at present there is no machine that approaches the kind of general-purpose human intelligence those applications would require. Despite their relative weakness on many fronts, human brains have some big advantages over their artificial competitors. Let me mention five:

- *Three-dimensionality:* Although, as noted, the linear dimensions of existing artificial processing units are vastly smaller than those of brains, the procedure by which they're made—centered on lithography (basically, etching)—is essentially two-dimensional. That is revealed visibly in the geometry of computer boards and chips. Of course, one can stack boards, but the spacing between layers is much larger, and communication much less efficient, than within layers. Brains make better use of all three dimensions.
- *Self-repair:* Human brains can recover from, or work around, many kinds of injuries or errors. Computers often must be repaired or rebooted externally.
- *Connectivity:* Human neurons typically support several hundred connections (synapses). Moreover, the complex pattern of these connections is very meaningful. (See my next point.) Computer units typically make only a handful of connections, in regular, fixed patterns.
- *Development* (self-assembly with interactive sculpting): The human brain grows its units by cell divisions and orchestrates them into coherent structures by movement and folding. It also proliferates an abundance of connections among the cells. An important part of its sculpting occurs through active processes during infancy and childhood, as the individual interacts with his or her environment. In

this process, many connections are winnowed away, while others are strengthened, depending on their effectiveness in use. Thus, the fine structure of the brain is tuned through interaction with the external world—a rich source of information and feedback!

- *Integration* (sensors and actuators): The human brain comes equipped with a variety of sensory organs, notably including its out-growth eyes, and with versatile actuators, including hands that build, legs that walk, and mouths that speak. Those sensors and actuators are seamlessly integrated into the brain's information-processing centers, having been honed over millions of years of natural selection. We interpret their raw signals and control their large-scale actions with minimal conscious attention. The flip side is that we don't know how we do it, and the implementation is opaque. It's proving surprisingly difficult to reach human standards on these "routine" input-output functions.

These advantages of human brains over currently engineered arti-facts are profound. Human brains supply an inspiring existence proof, showing us several ways we can get more out of matter. When, if ever, will our engineering catch up?

I don't know for sure, but let me offer some informed opinions. The challenges of three-dimensionality and, to a lesser extent, self-repair don't look overwhelming. They present some tough engineering prob-lems, but many incremental improvements are easy to imagine, and there are clear paths forward. And while the powers of human eyes, hands, and other sensory organs and actuators are wonderfully effective, their abilities are far from exhausting any physical limits. Optical systems can take pictures with higher resolution in space, time, and color, and in more regions of the electromagnetic spectrum; robots can move faster and be stronger; and so forth. In these domains, the components neces-sary for superhuman performance, along many axes, are already avail-able. The bottleneck is getting information into and out of them rapidly, in the language of the information-processing units.

And this brings us to the remaining, and I think most profound, advantages of brains over artificial devices, which stem from their connectivity and interactive development. Those two advantages are synergistic, since it is interactive development that sculpts the massively wired but sprawling structure of the infant brain, enabled by exponential growth of neurons and synapses, to get tuned in to the extraordinary instrument it becomes. Computer scientists are beginning to discover the power of the brain's architecture: Neural nets, whose basic design, as their name suggests, was directly inspired by the brain's, have scored some spectacular successes in game playing and pattern recognition, as noted. But present-day engineering has nothing comparable—in the (currently) esoteric domain of self-reproducing machines—to the power and versatility of neurons and their synapses. This could become a new, great frontier of research. Here, too, biology might point the way, as we come to understand biological development well enough to imitate its essence.

Altogether, the advantages of artificial over natural intelligence appear permanent, while the advantages of natural over artificial intelligence, though substantial at present, appear transient. I'd guess that it will be many decades before engineering catches up, but—barring catastrophic wars, climate change, or plagues, so that technological progress stays vigorous—few centuries.

If that's right, we can look forward to several generations during which humans, empowered and augmented by smart devices, coexist with increasingly capable autonomous AIs. There will be a complex, rapidly changing ecology of intelligence, and rapid evolution in consequence. Given the intrinsic advantages that engineered devices will eventually offer, the vanguard of that evolution will be cyborgs and superminds, rather than lightly adorned *Homo sapiens.*

Another important impetus will come from the exploration of hostile environments, both on Earth (e.g., the deep ocean) and, especially, in space. The human body is poorly adapted to conditions outside a narrow band of temperatures, pressures, and atmospheric composition. It needs a wide variety of specific, complex nutrients, and plenty of water. Also, it is not radiation hardened. As the manned space program has amply demonstrated, it is difficult and expensive to maintain humans outside their terrestrial comfort zone. Cyborgs or autonomous AIs could be much more effective in these explorations. Quantum AIs, with their sensitivity to noise, might even be happier in the cold and dark of deep space.

In a moving passage from his 1935 novel *Odd John,* science fiction's singular genius Olaf Stapledon has his hero, a superhuman (mutant) intelligence, describe *Homo sapiens* as "the Archaeopteryx of the spirit." He says this fondly to his friend and biographer, who is a normal human. Archaeopteryx was a noble creature, and a bridge to greater ones.

Chapter 8

LET'S ASPIRE TO MORE THAN MAKING OURSELVES OBSOLETE

MAX TEGMARK

Max Tegmark is an MIT physicist and AI researcher, president of the Future of Life Institute, scientific director of the Foundational Questions Institute, and the author of Our Mathematical Universe *and* Life 3.0: Being Human in the Age of Artificial Intelligence.

I was introduced to **Max Tegmark** some years ago by his MIT colleague Alan Guth, the father of inflation theory. A distinguished theoretical physicist and cosmologist himself, Max's principal concern nowadays is the looming existential risk posed by the creation of an AGI (artificial general intelligence—that is, one that matches human intelligence). Four years ago, Max co-founded, with Jaan Tallinn and others, the Future of Life Institute (FLI), which bills itself as "an outreach organization working to ensure that tomorrow's most powerful technologies are beneficial for humanity." While on a book tour in London, he was in the midst of planning for FLI, and he admits to being driven to tears in a tube station after a trip to the London Science Museum, with its exhibitions spanning the gamut of humanity's technological achievements. Was all that impressive progress in vain?

FLI's scientific advisory board includes Elon Musk, Frank Wilczek, George Church, Stuart Russell, and the Oxford philosopher Nick Bostrom, who dreamed up an oft-quoted Gedankenexperiment that results in a world full of paper clips and nothing else, produced by an (apparently) well-meaning AGI who was just following orders. The institute sponsors conferences (Puerto Rico 2015, Asilomar 2017) on AI safety issues and in 2018 instituted a grants competition focusing on research in aid of maximizing the societal benefits of AGI.

While Max is sometimes listed—by the noncognoscenti—on the side of the scaremongers, he believes, like Frank Wilczek, in a future that will immensely benefit from AGI if, in the attempt to create it, we can keep the human species from being sidelined.

lthough there's great controversy about how and when AI will impact humanity, the situation is clearer from a cosmic perspective: The technology-developing life that has evolved on Earth is rushing to make itself obsolete without devoting much serious thought to the consequences. This strikes me as embarrassingly lame, given that we can create amazing opportunities for humanity to flourish like never before, if we dare to steer a more ambitious course.

Our Universe has become aware of itself 13.8 billion years after its birth. On a small blue planet, tiny conscious parts of our Universe have discovered that what they once thought was the sum total of existence was a minute part of something far grander: a solar system in a galaxy in a universe with more than 100 billion other galaxies, arranged into an elaborate pattern of groups, clusters, and superclusters.

Consciousness is the cosmic awakening; it transformed our Universe from a mindless zombie with no self-awareness into a living ecosystem harboring self-reflection, beauty, hope, meaning, and purpose. Had that awakening never taken place, our Universe would have been pointless—a gigantic waste of space. Should our Universe go back to sleep permanently due to some cosmic calamity or self-inflicted mishap, it will become meaningless again.

On the other hand, things could get even better. We don't yet know whether we humans are the only stargazers in the cosmos, or even the first, but we've already learned enough about our Universe to know that it has the potential to wake up much more fully than it has thus far. AI pioneers such as Norbert Wiener have taught us that a further awakening of our Universe's ability to process and experience information need not require eons of additional evolution but perhaps mere decades of human scientific ingenuity.

We may be like that first glimmer of self-awareness you experienced when you emerged from sleep this morning, a premonition of the much greater consciousness that would arrive once you opened your eyes and fully awoke. Perhaps artificial superintelligence will enable life to spread throughout the cosmos and flourish for billions or trillions of years, and perhaps this will be because of decisions we make here, on our planet, in our lifetime.

Or humanity may soon go extinct, through some self-inflicted calamity caused by the power of our technology growing faster than the wisdom with which we manage it.

THE EVOLVING DEBATE ABOUT AI'S SOCIETAL IMPACT

Many thinkers dismiss the idea of superintelligence as science fiction, because they view intelligence as something mysterious that can exist only in biological organisms—especially humans—and as fundamentally limited to what today's humans can do. But from my perspective as a physicist, intelligence is simply a certain kind of information processing performed by elementary particles moving around, and there's no law of physics that says one can't build machines more intelligent in every way than we are, and able to seed cosmic life. This suggests that we've seen just the tip of the intelligence iceberg; there's an amazing potential to unlock the full intelligence latent in nature and use it to help humanity flourish—or flounder.

Others, including some of the authors in this volume, dismiss the building of an AGI (artificial general intelligence—an entity able to accomplish any cognitive task at least as well as humans) not because they consider it physically impossible but because they deem it too difficult for humans to pull off in less than a century. Among professional AI researchers, both types of dismissal have become minority views because of recent breakthroughs. There is a strong expectation that AGI will be achieved within a century, and the median forecast is only decades away. A recent survey of AI researchers by Vincent Müller and Nick Bostrom concludes:

> [T]he results reveal a view among experts that AI systems will probably (over 50%) reach overall human ability by 2040–50, and very likely (with 90% probability) by 2075. From reaching human ability, it will move on to superintelligence in 2 years (10%) to 30 years (75%) thereafter.*

In the cosmic perspective of gigayears, it makes little difference whether AGI arrives in thirty or three hundred years, so let's focus on the implications rather than the timing.

First, we humans discovered how to replicate some natural processes with machines, making our own heat, light, and mechanical horsepower. Gradually we realized that our bodies were also machines, and the discovery of nerve cells blurred the boundary between body and mind. Finally, we started building machines that could outperform not only our muscles but our minds as well. We've now been eclipsed by machines in the performance of many narrow cognitive tasks, ranging from memorization and arithmetic to game play, and we are in the process of being overtaken in many more, from driving to investing to medical diagnosing. If the AI community succeeds in its original goal of building AGI, then we will have, by definition, been eclipsed at all cognitive tasks.

This begs many obvious questions. For example, will whoever or whatever controls the AGI control Earth? Should we aim to control superintelligent machines? If not, can we ensure that they understand,

---

* Vincent C. Müller and Nick Bostrom, "Future Progress in Artificial Intelligence: A Survey of Expert Opinion," in *Fundamental Issues of Artificial Intelligence*, ed. Vincent C. Müller (Switzerland: Springer International Publishing, 2016), 555–72, https://nickbostrom.com/papers/survey.pdf.

adopt, and retain human values? As Norbert Wiener put it in *The Human Use of Human Beings:*

> Woe to us if we let [the machine] decide our conduct, unless we
> have previously examined the laws of its action, and know fully that its
> conduct will be carried out on principles acceptable to us! On the
> other hand, the machine . . . which can learn and can make decisions
> on the basis of its learning, will in no way be obliged to make such
> decisions as we should have made, or will be acceptable to us.

And who are the "us"? Who should deem "such decisions . . . acceptable"? Even if future powers decide to help humans survive and flourish, how will we find meaning and purpose in our lives if we aren't needed for anything?

The debate about the societal impact of AI has changed dramatically in the last few years. In 2014, what little public talk there was of AI risk tended to be dismissed as Luddite scaremongering, for one of two logically incompatible reasons:

1. AGI was overhyped and wouldn't happen for at least another century.
2. AGI would probably happen sooner but was virtually guaranteed to be beneficial.

Today, talk of AI's societal impact is everywhere, and work on AI safety and AI ethics has moved into companies, universities, and academic conferences. The controversial position on AI safety research is no longer to advocate for it but to dismiss it. Whereas the open letter that emerged from the 2015 Puerto Rico AI conference (and helped mainstream AI safety) spoke only in vague terms about the importance of keeping AI beneficial, the 2017 Asilomar AI Principles (see page 84) had real teeth: They explicitly mention recursive self-improvement, superintelligence, and existential risk, and were signed by AI industry leaders and more than a thousand AI researchers from around the world.

Nonetheless, most discussion is limited to the near-term impact of narrow AI and the broader community pays only limited attention to the dramatic transformations that AGI may soon bring to life on Earth. Why?

WHY WE'RE RUSHING TO MAKE OURSELVES OBSOLETE, AND WHY WE AVOID TALKING ABOUT IT

First of all, there's simple economics. Whenever we figure out how to make another type of human work obsolete by building machines that do it better and cheaper, most of society gains: Those who build and use the machines make profits, and consumers get more affordable products. This will be as true of future investor AGIs and scientist AGIs as it was of weaving machines, excavators, and industrial robots. In the past, displaced workers usually found new jobs, but this basic economic incentive will remain even if that is no longer the case. The existence of affordable AGI means, by definition, that *all* jobs can be done more cheaply by machines, so anyone claiming that "people will always find new well-paying jobs" is in effect claiming that AI researchers will fail to build AGI.

Second, *Homo sapiens* is by nature curious, which will motivate the scientific quest for understanding intelligence and developing AGI even without economic incentives. Although curiosity is one of the most celebrated human attributes, it can cause problems when it fosters technology we haven't yet learned how to manage wisely. Sheer scientific curiosity without profit motive contributed to the discovery of nuclear weapons and tools for engineering pandemics, so it's not unthinkable that the old adage "Curiosity killed the cat" will turn out to apply to the human species as well.

Third, we're mortal. This explains the near unanimous support for developing new technologies that help us live longer, healthier lives, which strongly motivates current AI research. AGI can clearly aid medical research even more. Some thinkers even aspire to near immortality via cyborgization or uploading.

We're thus on the slippery slope toward AGI, with strong incentives to keep sliding downward, even though the consequence will by definition be our economic obsolescence. We will no longer be needed for anything, because all jobs can be done more efficiently by machines. The successful creation of AGI would be the biggest event in human history, so why is there so little serious discussion of what it might lead to?

Here again, the answer involves multiple reasons.

First, as Upton Sinclair famously quipped, "It is difficult to get a man to understand something, when his salary depends on his not understanding it."* For example, spokesmen for tech companies or university research groups often claim there are no risks attached to their activities even if they privately think otherwise. Sinclair's observation may help explain not only reactions to risks from smoking and climate change but also why some treat technology as a new religion whose central articles of faith are that more technology is always better and whose heretics are clueless scaremongering Luddites.

Second, humans have a long track record of wishful thinking, flawed extrapolation of the past, and underestimation of emerging technologies. Darwinian evolution endowed us with powerful fear of concrete threats, not of abstract threats from future technologies that are hard to visualize or even imagine. Consider trying to warn people in 1930 of a future nuclear arms race, when you couldn't show them a single nuclear explosion video and nobody even knew how to build such weapons. Even top scientists can underestimate uncertainty, making forecasts that are either too optimistic—Where are those fusion reactors and flying cars?—or too pessimistic. Ernest Rutherford, arguably the greatest nuclear physicist of his time, said in 1933—less than twenty-four hours before Leo Szilard conceived of the nuclear chain reaction—that nuclear energy was "moonshine." Essentially nobody at that time saw the nuclear arms race coming.

Third, psychologists have discovered that we tend to avoid thinking of disturbing threats when we believe there's nothing we can do about

---

* Upton Sinclair, *I, Candidate for Governor: And How I Got Licked* (Berkeley: University of California Press, 1994), 109.

them anyway. In this case, however, there are many constructive things we can do, if we can get ourselves to start thinking about the issue.

WHAT CAN WE DO?

I'm advocating a strategy change from "Let's rush to build technology that makes us obsolete—what could possibly go wrong?" to "Let's envision an inspiring future and steer toward it."

To motivate the effort required for steering, this strategy begins by envisioning an enticing destination. Although Hollywood's futures tend to be dystopian, the fact is that AGI can help life flourish as never before. Everything I love about civilization is the product of intelligence, so if we can amplify our own intelligence with AGI, we have the potential to solve today's and tomorrow's thorniest problems, including disease, climate change, and poverty. The more detailed we can make our shared positive visions for the future, the more motivated we will be to work together to realize them.

What should we do in terms of steering? The twenty-three Asilomar principles adopted in 2017 offer plenty of guidance, including these short-term goals:

1. An arms race in lethal autonomous weapons should be avoided.
2. The economic prosperity created by AI should be shared broadly, to benefit all of humanity.
3. Investments in AI should be accompanied by funding for research on ensuring its beneficial use. . . . How can we make future AI systems highly robust, so that they do what we want without malfunctioning or getting hacked?*

The first two involve not getting stuck in suboptimal Nash equilibria. An out-of-control arms race in lethal autonomous weapons that drives

---

* https://futureoflife.org/ai-principles.

the price of automated anonymous assassination toward zero will be very hard to stop once it gains momentum. The second goal would require reversing the current trend in some Western countries where sectors of the population are getting poorer in absolute terms, fueling anger, resentment, and polarization. Unless the third goal can be met, all the wonderful AI technology we create might harm us, either accidentally or deliberately.

AI safety research must be carried out with a strict deadline in mind: Before AGI arrives, we need to figure out how to make AI understand, adopt, and retain our goals. The more intelligent and powerful machines get, the more important it becomes to align their goals with ours. As long as we build relatively dumb machines, the question isn't whether human goals will prevail but merely how much trouble the machines can cause before we solve the goal-alignment problem. If a superintelligence is ever unleashed, however, it will be the other way around: Since intelligence is the ability to accomplish goals, a superintelligent AI is by definition much better at accomplishing its goals than we humans are at accomplishing ours, and will therefore prevail.

In other words, *the real risk with AGI isn't malice but competence.* A superintelligent AGI will be extremely good at accomplishing its goals, and if those goals aren't aligned with ours, we're in trouble. People don't think twice about flooding anthills to build hydroelectric dams, so let's not place humanity in the position of those ants. Most researchers argue that if we end up creating superintelligence, we should make sure it's what AI-safety pioneer Eliezer Yudkowsky has termed "friendly AI"— AI whose goals are in some deep sense beneficial.

The moral question of what these goals should be is just as urgent as the technical questions about goal alignment. For example, what sort of society are we hoping to create, where we find meaning and purpose in our lives even though we, strictly speaking, aren't needed? I'm often given the following glib response to this question: *"Let's build machines that are smarter than us and then let them figure out the answer!"* This mistakenly equates intelligence with morality. Intelligence isn't good or evil but morally neutral. It's simply an ability to accomplish complex goals, good or bad. We can't conclude that things would have been better if Hitler had been more intelligent. Indeed, postponing work on ethical issues until after goal-aligned AGI is built would be irresponsible and potentially disastrous. A perfectly obedient superintelligence whose goals automatically align with those of its human owner would be like Nazi SS-Obersturmbannführer Adolf Eichmann on steroids. Lacking a moral compass or inhibitions of its own, it would, with ruthless efficiency, implement its owner's goals, whatever they might be.*

When I speak of the need to analyze technology risk, I'm sometimes accused of scaremongering. But here at MIT, where I work, we know that such risk analysis isn't scaremongering: It's *safety engineering.* Before the moon-landing mission, NASA systematically thought through everything that could possibly go wrong when putting astronauts on top of a 110-meter rocket full of highly flammable fuel and launching them to a place where nobody could help them—and there were lots of things that could go wrong. Was this scaremongering? No, this was the safety engineering that ensured the mission's success. Similarly, we should analyze what could go wrong with AI to ensure that it goes right.

OUTLOOK

In summary, if our technology outpaces the wisdom with which we manage it, it can lead to our extinction. It's already caused the extinction of from 20 to 50 percent of all species on Earth, by some estimates,† and it would be ironic if we're next in line. It would also be pathetic, given that the opportunities offered by AGI are literally astronomical, potentially enabling life to flourish for billions of years not only on Earth but also throughout much of our cosmos.

Instead of squandering this opportunity through unscientific risk denial and poor planning, let's be ambitious! *Homo sapiens* is inspiringly

---

* See, for example, Hannah Arendt, *Eichmann in Jerusalem: A Report on the Banality of Evil* (New York: Penguin Classics, 2006).
† See Elizabeth Kolbert, *The Sixth Extinction: An Unnatural History* (New York: Henry Holt, 2014).

ambitious, as reflected in William Ernest Henley's famous lines from "Invictus": "I am the master of my fate, / I am the captain of my soul." Rather than drifting like a rudderless ship toward our own obsolescence, let's take on and overcome the technical and societal challenges standing between us and a good high-tech future. What about the existential challenges related to morality, goals, and meaning? There's no meaning encoded in the laws of physics, so instead of passively waiting for our Universe to give meaning to us, let's acknowledge and celebrate that it's we conscious beings who give meaning to our Universe. Let's create our own meaning, based on something more profound than having jobs. AGI can enable us to finally become the masters of our own destiny. Let's make that destiny a truly inspiring one!

Chapter 9

DISSIDENT MESSAGES

JAAN TALLINN

Jaan Tallinn, a computer programmer, theoretical physicist, and investor, is a co-developer of Skype and Kazaa.

Jaan Tallinn grew up in Estonia, becoming one of its few computer game developers, when that nation was still a Soviet Socialist Republic. Here he compares the dissidents who brought down the Iron Curtain to the dissidents who are sounding the alarm about rapid advances in artificial intelligence. He locates the roots of the current AI dissidence, paradoxically, among such pioneers of the AI field as Wiener, Alan Turing, and I. J. Good.

Jaan's preoccupation is with existential risk, AI being among the most extreme of many. In 2012, he co-founded the Centre for the Study of Existential Risk—an interdisciplinary research institute that works to mitigate risks "associated with emerging technologies and human activity"—at the University of Cambridge, along with philosopher Huw Price and Martin Rees, the Astronomer Royal.

He once described himself to me as "a convinced con-sequentialist"—convinced enough to have given away much of his entrepreneurial wealth to the Future of Life Institute (of which he is a co-founder), the Machine Intelligence Research Institute, and other such organizations working on risk reduction. Max Tegmark has written about him: "If you're an intelligent life-form reading this text millions of years from now and marveling at how life is flourishing, you may owe your existence to Jaan."

On a recent visit to London, Jaan and I participated in an AI panel for the Serpentine Gallery's Marathon at London's City Hall, under the aegis of Hans Ulrich Obrist (another contributor to this volume). This being the art world, there was a glamorous dinner party that night in a mansion filled with London's beautiful people—artists, fashion models, oligarchs, stars of stage and screen. After working the room in his unaffected manner ("Hi, I'm Jaan"), he suddenly said, "Time for hip-hop dancing," dropped to the floor on one hand, and began demonstrating his spectacular moves to the bemused A-listers. Then off he went into the dance-club subculture, which is apparently how he ends every evening when he's on the road. Who knew?

n March 2009, I found myself in a bland franchise eatery next to a noisy California freeway. I was there to meet a young man whose blog I had been following. To make himself recognizable, he wore a button with a text on it: *Speak the truth even if your voice trembles.* His name was Eliezer Yudkowsky, and we spent the next four hours discussing the message he had for the world—a message that had brought me to that eatery and would end up dominating my subsequent work.

THE FIRST MESSAGE: THE SOVIET OCCUPATION

In *The Human Use of Human Beings,* Norbert Wiener looked at the world through the lens of communication. He saw a universe that was marching to the tune of the second law of thermodynamics toward its inevitable heat death. In such a universe, the only (meta)stable entities are *messages*—patterns of information that propagate through time, like waves propagating across the surface of a lake. Even we humans can be considered messages, because the atoms in our bodies are too fleeting to attach our identities to. Instead, we are the "message" that our bodily functions maintain. As Wiener put it: "It is the pattern maintained by this homeostasis, which is the touchstone of our personal identity."

I'm more used to treating processes and computation as the

fundamental building blocks of the world. That said, Wiener's lens brings out some interesting aspects of the world that might otherwise have remained in the background and that to a large degree shaped my life. These are two messages, both of which have their roots in the Second World War. They started out as quiet dissident messages—messages that people didn't pay much attention to, even if they silently and perhaps subconsciously concurred. The first message was: *The Soviet Union is composed of a series of illegitimate occupations. These occupations must end.*

As an Estonian, I grew up behind the Iron Curtain and had a front-row seat when it fell. I heard this first message in the nostalgic reminiscences of my grandparents and in between the harsh noises jamming the Voice of America. It grew louder during the Gorbachev era, as the state became more lenient in its treatment of dissidents, and reached a crescendo in the Estonian Singing Revolution of the late 1980s.

In my teens, I witnessed the message spread out across widening circles of people, starting with the active dissidents, who had voiced it for half a century at great cost to themselves, proceeding to the artists and literati, and ending up among the party members and politicians who had switched sides. This new elite comprised an eclectic mix of people: those original dissidents who had managed to survive the repression, public intellectuals, and (to the great annoyance of the surviving dissidents) even former Communists. The remaining dogmatists—even the prominent ones—were eventually marginalized, some of them retreating to Russia.

Interestingly, as the message propagated from one group to the next, it evolved. It started in pure and uncompromising form ("The occupation must end!") among the dissidents who considered the truth more important than their personal freedom. The mainstream groups, who had more to lose, initially qualified and diluted the message, taking positions like "It would make sense in the long term to delegate control over local matters." (There were always exceptions: Some public intellectuals proclaimed the original dissident message verbatim.) Finally, the original message—being, simply, true—won out over its diluted versions.

Estonia regained its independence in 1991, and the last Soviet troops left three years later.

The people who took the risk and spoke the truth in Estonia and elsewhere in the Eastern Bloc played a monumental role in the eventual outcome—an outcome that changed the lives of hundreds of millions of people, myself included. They spoke the truth, even as their voices trembled.

THE SECOND MESSAGE: AI RISK

My exposure to the second revolutionary message was via Yudkowsky's blog—the blog that compelled me to reach out and arrange that meeting in California. The message was: *Continued progress in AI can precipitate a change of cosmic proportions—a runaway process that will likely kill everyone. We need to put in a lot of extra effort to avoid that outcome.*

After my meeting with Yudkowsky, the first thing I did was try to interest my Skype colleagues and close collaborators in his warning. I failed. The message was too crazy, too dissident. Its time had not yet come.

Only later did I learn that Yudkowsky wasn't the original dissident speaking this particular truth. In April 2000, there was a lengthy opinion piece in *Wired* titled "Why the Future Doesn't Need Us" by Bill Joy, co-founder and chief scientist of Sun Microsystems. He warned:

> Accustomed to living with almost routine scientific break-throughs, we have yet to come to terms with the fact that the most compelling 21st-century technologies—robotics, genetic engineering, and nanotechnology—pose a different threat than the technologies that have come before. Specifically, robots, engineered organisms, and nanobots share a dangerous amplifying factor: They can self-replicate. . . . [O]ne bot can become many, and quickly get out of control.

Apparently, Joy's broadside caused a lot of furor but little action.

More surprising to me, though, was that the AI-risk message arose almost simultaneously with the field of computer science. In a 1951 lecture, Alan Turing announced: "[I]t seems probable that once the machine thinking method had started, it would not take long to outstrip our feeble powers. . . . At some stage, therefore, we should have to expect the machines to take control. . . ."* A decade or so later, his Bletchley Park colleague I. J. Good wrote, "The first ultraintelligent machine is the *last* invention that man need ever make, provided that the machine is docile enough to tell us how to keep it under control."† Indeed, I counted half a dozen places in *The Human Use of Human Beings* where Wiener hinted at one or another aspect of the Control Problem. ("The machine like the djinnee, which can learn and can make decisions on the basis of its learning, will in no way be obliged to make such decisions as we should have made, or will be acceptable to us.") Apparently, the original dissidents promulgating the AI-risk message were the AI pioneers themselves!

EVOLUTION'S FATAL MISTAKE

There have been many arguments, some sophisticated and some less so, for why the Control Problem is real and not some science-fiction fantasy. Allow me to offer one that illustrates the magnitude of the problem:

For the last hundred thousand years, the world (meaning the Earth, but the argument extends to the solar system and possibly even to the entire universe) has been in the human-brain regime. In this regime, the brains of *Homo sapiens* have been the most sophisticated future-shaping mechanisms (indeed, some have called them the most complicated objects in the universe). Initially, we didn't use them for much beyond

* Posthumously reprinted in *Philosophia Mathematica* 4, no. 3 (1966): 256–60.
† Irving John Good, "Speculations Concerning the First Ultraintelligent Machine," *Advances in Computers* 6 (Cambridge, MA: Academic Press, 1965): 31–88.

survival and tribal politics in a band of foragers, but now their effects are surpassing those of natural evolution. The planet has gone from producing forests to producing cities.

As predicted by Turing, once we have superhuman AI ("the machine thinking method"), the human-brain regime will end. Look around you—you're witnessing the final decades of a hundred-thousand-year regime. This thought alone should give people some pause before they dismiss AI as just another tool. One of the world's leading AI researchers recently confessed to me that he would be greatly relieved to learn that human-level AI was impossible for us to create.

Of course, it might still take us a long time to develop human-level AI. But we have reason to suspect that this is not the case. After all, it didn't take long, in relative terms, for evolution—the blind and clumsy optimization process—to create human-level intelligence once it had animals to work with. Or multicellular life, for that matter: Getting cells to stick together seems to have been much harder for evolution to accomplish than creating humans once there were multicellular organisms. Not to mention that our level of intelligence was limited by such grotesque factors as the width of the birth canal. Imagine an AI developer being stopped in his tracks because he couldn't manage to adjust the font size on his computer!

There's an interesting symmetry here: In fashioning humans, evolution created a system that is, at least in many important dimensions, a more powerful planner and optimizer than evolution itself is. We are the first species to understand that we're the product of evolution. Moreover, we've created many artifacts (radios, firearms, spaceships) that evolution would have little hope of creating. Our future, therefore, will be determined by our own decisions and no longer by biological evolution. In that sense, evolution has fallen victim to its own Control Problem.

We can only hope that we're smarter than evolution in that sense. We *are* smarter, of course, but will that be enough? We're about to find out.

THE PRESENT SITUATION

So here we are, more than half a century after the original warnings by Turing, Wiener, and Good, and a decade after people like me started paying attention to the AI-risk message. I'm glad to see that we've made a lot of progress in confronting this issue, but we're definitely not there yet. AI risk, although no longer a taboo topic, is not yet fully appreciated among AI researchers. AI risk is not yet common knowledge either. In relation to the timeline of the first dissident message, I'd say we're around the year 1988, when raising the Soviet-occupation topic was no longer a career-ending move but you still had to somewhat hedge your position. I hear similar hedging now—statements like "I'm not concerned about superintelligent AI, but there are some real ethical issues in increased automation," or "It's good that some people are researching AI risk, but it's not a short-term concern," or even the very reasonable-sounding "These are small-probability scenarios, but their potentially high impact justifies the attention."

As far as message propagation goes, though, we are getting close to the tipping point. A recent survey of AI researchers who published at the two major international AI conferences in 2015 found that 40 percent now think that risks from highly advanced AI are either "an important problem" or "among the most important problems in the field."*

Of course, just as there were dogmatic Communists who never changed their position, it's all but guaranteed that some people will never admit that AI is potentially dangerous. Many of the deniers of the first kind came from the Soviet nomenklatura; similarly, the AI-risk deniers often have financial or other pragmatic motives. One of the leading motives is corporate profits. AI is profitable, and even in instances where it isn't, it's at least a trendy, forward-looking enterprise with which to associate your company. So a lot of the dismissive positions are products of corporate PR and legal machinery. In some very real sense, big

---

* Katja Grace et al., "When Will AI Exceed Human Performance? Evidence from AI Experts," https://arxiv .org/pdf/1705.08807.pdf.

corporations are nonhuman machines that pursue their own interests—interests that might not align with those of any particular human working for them. As Wiener observed in *The Human Use of Human Beings*: "When human atoms are knit into an organization in which they are used, not in their full right as responsible human beings, but as cogs and levers and rods, it matters little that their raw material is flesh and blood."

Another strong incentive to turn a blind eye to the AI risk is the (very human) curiosity that knows no bounds. "When you see something that is technically sweet, you go ahead and do it and you argue about what to do about it only after you have had your technical success. That is the way it was with the atomic bomb," said J. Robert Oppenheimer. His words were echoed recently by Geoffrey Hinton, arguably the inventor of deep learning, in the context of AI risk: "I could give you the usual arguments, but the truth is that the prospect of discovery is too sweet."

Undeniably, we have both entrepreneurial attitude and scientific curiosity to thank for almost all the nice things we take for granted in the modern era. It's important to realize, though, that progress does not *owe* us a good future. In Wiener's words, "It is possible to believe in progress as a fact without believing in progress as an ethical principle."

Ultimately, we don't have the luxury of waiting before all the corporate heads and AI researchers are willing to concede the AI risk. Imagine yourself sitting in a plane about to take off. Suddenly there's an announcement that 40 percent of the experts believe there's a bomb on board. At that point, the course of action is already clear, and sitting there waiting for the remaining 60 percent to come around isn't part of it.

CALIBRATING THE AI-RISK MESSAGE

While uncannily prescient, the AI-risk message from the original dissidents has a giant flaw—as does the version dominating current public discourse: Both considerably understate the magnitude of the problem as well as AI's potential upside. The message, in other words, does not adequately convey the stakes of the game.

Wiener primarily warned of the *social* risks—risks stemming from careless integration of machine-generated decisions with governance processes and misuse (by humans) of such automated decision making. Likewise, the current "serious" debate about AI risks focuses mostly on things like technological unemployment or biases in machine learning. While such discussions can be valuable and address pressing short-term problems, they are also stunningly parochial. I'm reminded of Yudkowsky's quip in a blog post: "[A]sking about the effect of *machine superintelligence* on the conventional human labor market is like asking how US-Chinese trade patterns would be affected by the Moon crashing into the Earth. There would indeed be effects, but you'd be missing the point."

In my view, the central point of the AI risk is that *superintelligent AI is an environmental risk*. Allow me to explain.

In his parable of the "Sentient Puddle," Douglas Adams describes a puddle that wakes up in the morning and finds himself in a hole that fits him "staggeringly well." From that observation, the puddle concludes that the world must have been made for him. Therefore, writes Adams, "the moment he disappears catches him rather by surprise." To assume that AI risks are limited to adverse social developments is to make a similar mistake. The harsh reality is that the universe was not made for us; instead, we are fine-tuned by evolution to a very narrow range of environmental parameters. For instance, we need the atmosphere at ground level to be roughly at room temperature, at about 100 kPa pressure, and have a sufficient concentration of oxygen. Any disturbance, even temporary, of this precarious equilibrium and we die in a matter of minutes.

Silicon-based intelligence does not share such concerns about the environment. That's why it's much cheaper to explore space using machine probes rather than "cans of meat." Moreover, Earth's current environment is almost certainly suboptimal for what a superintelligent AI will greatly care about: *efficient computation*. Hence, we might find our

planet suddenly going from anthropogenic global warming to machino-
genic global cooling. One big challenge that AI safety research needs to
deal with is how to constrain a potentially superintelligent AI—an AI
with a much larger footprint than our own—from rendering our environment uninhabitable for biological life-forms.

Interestingly, given that the most potent sources of both AI research
and AI-risk dismissals are under big corporate umbrellas, if you squint
hard enough, the "AI as an environmental risk" message looks like the
chronic concern about corporations skirting their environmental
responsibilities.

Conversely, the worry about AI's social effects also misses most of the
upside. It's hard to overemphasize how tiny and parochial the future of
our planet is, compared with the full potential of humanity. On astronomical timescales, our planet will be gone soon (unless we tame the
sun, also a distinct possibility) and almost all the resources—atoms and
free energy—to sustain civilization in the long run are in deep space.

Eric Drexler, the inventor of nanotechnology, has recently been pop-
ularizing the concept of "Pareto-topia": the idea that AI, if done right,
can bring about a future in which *everyone's* lives are hugely improved,
a future where there are no losers. A key realization here is that what
chiefly prevents humanity from achieving its full potential might be our
instinctive sense that we're in a zero-sum game—a game in which players are supposed to eke out small wins at the expense of others. Such an
instinct is seriously misguided and destructive in a "game" where everything is at stake and the payoff is literally astronomical. There are many
more star systems in our galaxy alone than there are people on Earth.

HOPE

As of this writing, I'm cautiously optimistic that the AI-risk message can
save humanity from extinction, just as the Soviet-occupation message
ended up liberating hundreds of millions of people. As of 2015, it had
reached and converted 40 percent of AI researchers. It wouldn't surprise

me if a new survey now would show that the majority of AI researchers believe AI safety to be an important issue.

I'm delighted to see the first technical AI-safety papers coming out of DeepMind, OpenAI, and Google Brain and the collaborative problem-solving spirit flourishing among the AI-safety research teams in these otherwise very competitive organizations.

The world's political and business elite are also slowly waking up: AI safety has been covered in reports and presentations by the Institute of Electrical and Electronics Engineers (IEEE), the World Economic Forum, and the Organization for Economic Cooperation and Development (OECD). Even the recent (July 2017) Chinese AI manifesto contained dedicated sections on "AI safety supervision" and "Develop[ing] laws, regulations, and ethical norms" and establishing "an AI security and evaluation system" to, among other things, "[e]nhance the awareness of risk." I very much hope that a new generation of leaders who understand the AI Control Problem and AI as the ultimate environmental risk can rise above the usual tribal, zero-sum games and steer humanity past these dangerous waters we are in—thereby opening our way to the stars that have been waiting for us for billions of years.

Here's to our next hundred thousand years! And don't hesitate to speak the truth, even if your voice trembles.

Chapter 10

TECH PROPHECY AND THE UNDERAPPRECIATED CAUSAL POWER OF IDEAS

STEVEN PINKER

Steven Pinker, a Johnstone Family Professor in the Department of Psychology at Harvard University, is an experimental psychologist who conducts research in visual cognition, psycholinguistics, and social relations. He is the author of eleven books, including The Blank Slate, The Better Angels of Our Nature, *and, most recently,* Enlightenment Now: The Case for Reason, Science, Humanism, and Progress.

Throughout his career, whether studying language, advocating a realistic biology of mind, or examining the human condition through the lens of humanistic Enlightenment ideas, psychologist **Steven Pinker** has embraced and championed a naturalistic understanding of the universe and the computational theory of mind. He is perhaps the first internationally recognized public intellectual whose recognition is based on the advocacy of empirically based thinking about language, mind, and human nature.

"Just as Darwin made it possible for a thoughtful observer of the natural world to do without creationism," he says, "Turing and others made it possible for a thoughtful observer of the cognitive world to do without spiritualism." In the debate about AI risk, he argues against prophecies of doom and gloom, noting that they spring from the worst of our psychological biases—exemplified particularly by media reports: "Disaster scenarios are cheap to play out in the probability-free zone of our imaginations, and they can always find a worried, technophobic, or morbidly fascinated audience." Hence, over the centuries: Pandora, Faust, the Sorcerer's Apprentice, Frankenstein, the population bomb, resource depletion, HAL, suitcase nukes, the Y2K bug, and engulfment by nanotechnological grey goo. "A characteristic of AI dystopias," he points out, "is that they project a parochial alpha-male psychology onto the concept of intelligence. . . . History does turn up the occasional megalomaniacal despot or psychopathic serial killer, but these are products of a history of natural selection shaping testosterone-sensitive circuits in a certain species of primate, not an inevitable feature of intelligent systems."

In the present essay, he applauds Wiener's belief in the strength of ideas vis-à-vis the encroachment of technology. As Wiener so aptly put it, "The machine's danger to society is not from the machine itself but from what man makes of it."

A rtificial intelligence is an existence proof of one of the great ideas in human history: that the abstract realm of knowledge, reason, and purpose does not consist of an élan vital or immaterial soul or miraculous powers of neural tissue. Rather, it can be linked to the physical realm of animals and machines via the concepts of information, computation, and control. Knowledge can be explained as patterns in matter or energy that stand in systematic relations with states of the world, with mathematical and logical truths, and with one another. Reasoning can be explained as transformations of that knowledge by physical operations that are designed to preserve those relations. Purpose can be explained as the control of operations to effect changes in the world, guided by discrepancies between its current state and a goal state. Naturally evolved brains are just the most familiar systems that achieve intelligence through information, computation, and control. Humanly designed systems that achieve intelligence vindicate the notion that information processing is sufficient to explain it—the notion that the late Jerry Fodor dubbed the computational theory of mind.

The touchstone for this volume, Norbert Wiener's *The Human Use of Human Beings,* celebrated this intellectual accomplishment, of which Wiener himself was a foundational contributor. A potted history of the mid-20th-century revolution that gave the world the computational theory of mind might credit Claude Shannon and Warren Weaver for explaining knowledge and communication in terms of information. It might credit Alan Turing and John von Neumann for explaining intelligence and reasoning in terms of computation. And it ought to give Wiener credit for explaining the hitherto mysterious world of purposes, goals, and teleology in terms of the technical concepts of feedback, control, and cybernetics (in its original sense of "governing" the operation of a goal-directed system). "It is my thesis," he announced, "that the physical functioning of the living individual and the operation of some of the newer communication machines are precisely parallel in their analogous attempts to control entropy through feedback"—the staving off of life-sapping entropy being the ultimate goal of human beings.

Wiener applied the ideas of cybernetics to a third system: society. The laws, norms, customs, media, forums, and institutions of a complex community could be considered channels of information propagation and feedback that allow a society to ward off disorder and pursue certain goals. This is a thread that runs through the book and which Wiener himself may have seen as its principal contribution. In his explanation of feedback, he wrote, "This complex of behavior is ignored by the average man, and in particular does not play the role that it should in our habitual analysis of society; for just as individual physical responses may be seen from this point of view, so may the organic responses of society itself."

Indeed, Wiener gave scientific teeth to the idea that in the workings of history, politics, and society, *ideas matter*. Beliefs, ideologies, norms, laws, and customs, by regulating the behavior of the humans who share them, can shape a society and power the course of historical events as surely as the phenomena of physics affect the structure and evolution of the solar system. To say that ideas—and not just weather, resources, geography, or weaponry—can shape history is not woolly mysticism. It is a statement of the causal powers of information instantiated in human brains and exchanged in networks of communication and feedback. Deterministic theories of history, whether they identify the causal engine as technological, climatological, or geographic, are belied by the causal power of ideas. The effects of these ideas can include unpredictable lurches and oscillations that arise from positive feedback or from mis-calibrated negative feedback.

An analysis of society in terms of its propagation of ideas also gave Wiener a guideline for social criticism. A healthy society—one that gives its members the means to pursue life in defiance of entropy—allows in-formation sensed and contributed by its members to feed back and af-fect how the society is governed. A dysfunctional society invokes dogma and authority to impose control from the top down. Wiener thus de-scribed himself as "a participant in a liberal outlook," and devoted most of the moral and rhetorical energy in the book (both the 1950 and 1954 editions) to denouncing communism, fascism, McCarthyism, militarism, and authoritarian religion (particularly Catholicism and Islam) and to warning that political and scientific institutions were becoming too hier-archical and insular.

Wiener's book is also, here and there, an early exemplar of an in-creasingly popular genre, tech prophecy. Prophecy not in the sense of mere prognostications but in the Old Testament sense of dark warnings of catastrophic payback for the decadence of one's contemporaries. Wie-ner warned against the accelerating nuclear arms race, against techno-logical change that was imposed without regard to human welfare ("[W]e must know as scientists what man's nature is and what his built-in pur-poses are"), and against what today is called the value-alignment prob-lem: that "the machine like the djinnee, which can learn and can make decisions on the basis of its learning, will in no way be obliged to make such decisions as we should have made, or will be acceptable to us." In the darker 1950 edition he warned of a "threatening new Fascism depen-dent on the *machine à gouverner.*"

Wiener's tech prophecy harks back to the Romantic movement's re-bellion against the "dark Satanic mills" of the Industrial Revolution, and perhaps even earlier, to the archetypes of Prometheus, Pandora, and Faust. And today it has gone into high gear. Jeremiahs, many of them (like Wiener) from the worlds of science and technology, have sounded alarms about nanotechnology, genetic engineering, Big Data, and particularly artificial intelligence. Several contributors to this volume characterize Wiener's book as a prescient example of tech prophecy and amplify his dire worries.

Yet the two moral themes of *The Human Use of Human Beings*—the liberal defense of an open society and the dystopian dread of runaway technology—are in tension. A society with channels of feedback that maximize human flourishing will have mechanisms in place, and can adapt them to changing circumstances, in a way that can domesticate technology to human purposes. There's nothing idealistic or mystical about this; as Wiener emphasized, ideas, norms, and institutions are themselves a form of technology, consisting of patterns of information distributed across brains. The possibility that machines threaten a new fascism must be weighed against the vigor of the liberal ideas, institutions, and norms that Wiener championed throughout the book. The flaw in today's dystopian prophecies is that they disregard the existence of these norms and institutions, or drastically underestimate their causal potency. The result is a technological determinism whose dark predictions are repeatedly refuted by the course of events. The numbers "1984" and "2001" are good reminders.

I will consider two examples. Tech prophets often warn of a "surveillance state" in which a government empowered by technology will monitor and interpret all private communications, allowing it to detect dissent and subversion as it arises and make resistance to state power futile. Orwell's telescreens are the prototype, and in 1976 Joseph Weizenbaum, one of the gloomiest tech prophets of all time, warned my class of graduate students not to pursue automatic speech recognition because government surveillance was its only conceivable application.

Though I am on record as an outspoken civil libertarian, deeply concerned with contemporary threats to free speech, I lose no sleep over technological advances in the Internet, video, or artificial intelligence. The reason is that almost all the variation across time and space in freedom of thought is driven by differences in norms and institutions and almost none of it by differences in technology. Though one can imagine hypothetical combinations of the most malevolent totalitarians with the most advanced technology, in the real world it's the norms and laws we should be vigilant about, not the tech.

Consider variation across time. If, as Orwell hinted, advancing technology was a prime enabler of political repression, then Western societies should have gotten more and more restrictive of speech over the centuries, with a dramatic worsening in the second half of the 20th century continuing into the 21st. That's not how history unfolded. It was the centuries when communication was implemented by quills and inkwells that had autos-da-fé and the jailing or guillotining of Enlightenment thinkers. During World War I, when the state of the art was the wireless, Bertrand Russell was jailed for his pacifist opinions. In the 1950s, when computers were room-size accounting machines, hundreds of liberal writers and scholars were professionally punished. Yet in the technologically accelerating, hyperconnected 21st century, 18 percent of social science professors are Marxists;* the president of the United States is nightly ridiculed by television comedians as a racist, pervert, and moron; and technology's biggest threat to political discourse comes from amplifying too many dubious voices rather than suppressing enlightened ones.

Now consider variations across place. Western countries at the technological frontier consistently get the highest scores in indexes of democracy and human rights, while many backward strongman states are at the bottom, routinely jailing or killing government critics. The lack of a correlation between technology and repression is unsurprising when you analyze the channels of information flow in any human society. For dissidents to be influential, they have to get their message out to a wide network via whatever channels of communication are available— pamphleteering, soap-box oration, subversive soirees in cafés and pubs, word of mouth. These channels enmesh influential dissidents in a broad social network, which makes them easy to identify and track down. All the more so when dictators rediscover the time-honored technique of

---

* Neil Gross and Solon Simmons, "The Social and Political Views of American College and University Professors," in *Professors and Their Politics*, ed. N. Gross and S. Simmons (Baltimore: Johns Hopkins University Press, 2014).

weaponizing the people against one another by punishing those who don't denounce or punish others.

In contrast, technologically advanced societies have long had the means to install Internet-connected, government-monitored surveillance cameras in every bar and bedroom. Yet that has not happened, because democratic governments (even the current American administration, with its flagrantly antidemocratic impulses) lack the will and the means to enforce such surveillance on an obstreperous people accustomed to saying what they want. Occasionally, warnings of nuclear, biological, or cyberterrorism goad government security agencies into measures such as hoovering up mobile phone metadata, but these ineffectual measures, more theater than oppression, have had no significant effect on either security or freedom. Ironically, tech prophecy plays a role in encouraging these measures. By sowing panic about supposed existential threats such as suitcase nuclear bombs and bioweapons assembled in teenagers' bedrooms, they put pressure on governments to prove they're doing something, anything, to protect the American people.

It's not that political freedom takes care of itself. It's that the biggest threats lie in the networks of ideas, norms, and institutions that allow information to feed back (or not) on collective decisions and understanding. As opposed to the chimerical technological threats, one real threat today is oppressive political correctness, which has choked the range of publicly expressible hypotheses, terrified many intelligent people against entering the intellectual arena, and triggered a reactionary backlash. Another real threat is the combination of prosecutorial discretion with an expansive lawbook filled with vague statutes. The result is that every American unwittingly commits "three felonies a day" (as the title of a book by civil libertarian Harvey Silverglate puts it) and is in jeopardy of imprisonment whenever it suits the government's needs. It's this prosecutorial weaponry that makes Big Brother all-powerful, not telescreens. The activism and polemicizing directed against government surveillance programs would be better directed at its overweening legal powers.

The other focus of much tech prophecy today is artificial intelligence, whether in the original sci-fi dystopia of computers running amok and enslaving us in an unstoppable quest for domination or the newer version, in which they subjugate us by accident, single-mindedly seeking some goal we give them regardless of its side effects on human welfare (the value-alignment problem adumbrated by Wiener). Here again both threats strike me as chimerical, growing from a narrow technological determinism that neglects the networks of information and control in an intelligent system, like a computer or a brain, and in a society as a whole.

The subjugation fear is based on a muzzy conception of intelligence that owes more to the Great Chain of Being and a Nietzschean will to power than to a Wienerian analysis of intelligence and purpose in terms of information, computation, and control. In these horror scenarios, intelligence is portrayed as an all-powerful, wish-granting potion that agents possess in different amounts. Humans have more of it than animals, and an artificially intelligent computer or robot will have more of it than humans. Since we humans have used our moderate endowment to domesticate or exterminate less well-endowed animals (and since technologically advanced societies have enslaved or annihilated technologically primitive ones), it follows that a supersmart AI would do the same to us. Since an AI will think millions of times faster than we do, and use its superintelligence to recursively improve its superintelligence, from the instant it is turned on we will be powerless to stop it.

But these scenarios are based on a confusion of intelligence with motivation—of beliefs with desires, inferences with goals, the computation elucidated by Turing and the control elucidated by Wiener. Even if we did invent superhumanly intelligent robots, why would they *want* to enslave their masters or take over the world? Intelligence is the ability to deploy novel means to attain a goal. But the goals are extraneous to the intelligence: Being smart is not the same as wanting something. It just so happens that the intelligence in *Homo sapiens* is a product of Darwinian natural selection, an inherently competitive process. In the brains of that species, reasoning comes bundled with goals such as dominating rivals and amassing resources. But it's a mistake to confuse a circuit in the limbic brain of a certain species of primate with the very nature of intelligence. There is no law of complex systems that says that intelligent agents must turn into ruthless megalomaniacs.

A second misconception is to think of intelligence as a boundless continuum of potency, a miraculous elixir with the power to solve any problem, attain any goal. The fallacy leads to nonsensical questions like when an AI will "exceed human-level intelligence," and to the image of an "artificial general intelligence" (AGI) with God-like omniscience and omnipotence. Intelligence is a contraption of gadgets: software modules that acquire, or are programmed with, knowledge of how to pursue various goals in various domains. People are equipped to find food, win friends and influence people, charm prospective mates, bring up children, move around in the world, and pursue other human obsessions and pastimes. Computers may be programmed to take on some of these problems (like recognizing faces), not to bother with others (like charming mates), and to take on still other problems that humans can't solve (like simulating the climate or sorting millions of accounting records). The problems are different, and the kinds of knowledge needed to solve them are different.

But instead of acknowledging the centrality of knowledge to intelligence, the dystopian scenarios confuse an artificial general intelligence of the future with Laplace's demon, the mythical being that knows the location and momentum of every particle in the universe and feeds them into equations for physical laws to calculate the state of everything at any time in the future. For many reasons, Laplace's demon will never be implemented in silicon. A real-life intelligent system has to acquire information about the messy world of objects and people by engaging with it one domain at a time, the cycle being governed by the pace at which events unfold in the physical world. That's one of the reasons that understanding does not obey Moore's Law: Knowledge is acquired by formulating explanations and testing them against reality, not by running an algorithm faster and faster. Devouring the information on the Internet will not confer omniscience either: Big Data is still finite data, and the universe of knowledge is infinite.

A third reason to be skeptical of a sudden AI takeover is that it takes too seriously the inflationary phase in the AI hype cycle in which we are living today. Despite the progress in machine learning, particularly multilayered artificial neural networks, current AI systems are nowhere near achieving general intelligence (if that concept is even coherent). Instead, they are restricted to problems that consist of mapping well-defined inputs to well-defined outputs in domains where gargantuan training sets are available, in which the metric for success is immediate and precise, in which the environment doesn't change, and in which no stepwise, hierarchical, or abstract reasoning is necessary. Many of the successes come not from a better understanding of the workings of intelligence but from the brute-force power of faster chips and Bigger Data, which allow the programs to be trained on millions of examples and generalize to similar new ones. Each system is an idiot savant, with little ability to leap to problems it was not set up to solve and a brittle mastery of those it was. And to state the obvious, none of these programs has made a move toward taking over the lab or enslaving its programmers.

Even if an artificial intelligence system tried to exercise a will to power, without the cooperation of humans it would remain an impotent brain in a vat. A superintelligent system, in its drive for self-improvement, would somehow have to build the faster processors that it would run on, the infrastructure that feeds it, and the robotic effectors that connect it to the world—all impossible unless its human victims worked to give it control of vast portions of the engineered world. Of course, one can always imagine a Doomsday Computer that is malevolent, universally empowered, always on, and tamperproof. The way to deal with this threat is straightforward: Don't build one.

What about the newer AI threat, the value-alignment problem, foreshadowed in Wiener's allusions to stories of the monkey's paw, the genie, and King Midas, in which a wisher rues the unforeseen side effects of his wish? The fear is that we might give an AI system a goal and then helplessly stand by as it relentlessly and literal-mindedly implemented its interpretation of that goal, the rest of our interests be damned. If we gave an AI the goal of maintaining the water level behind a dam, it might flood a town, not caring about the people who drowned. If we gave it the goal of making paper clips, it might turn all the matter in the reachable universe into paper clips, including our possessions and bodies. If we asked it to maximize human happiness, it might implant us all with intravenous dopamine drips, or rewire our brains so we were happiest sitting in jars, or, if it had been trained on the concept of happiness with pictures of smiling faces, tile the galaxy with trillions of nanoscopic pictures of smiley faces.

Fortunately, these scenarios are self-refuting. They depend on the premises that (1) humans are so gifted that they can design an omniscient and omnipotent AI, yet so idiotic that they would give it control of the universe without testing how it works; and (2) the AI would be so brilliant that it could figure out how to transmute elements and rewire brains, yet so imbecilic that it would wreak havoc based on elementary blunders of misunderstanding. The ability to choose an action that best satisfies conflicting goals is not an add-on to intelligence that engineers might forget to install and test; it *is* intelligence. So is the ability to interpret the intentions of a language user in context.

When we put aside fantasies like digital megalomania, instant omniscience, and perfect knowledge and control of every particle in the universe, artificial intelligence is like any other technology. It is developed incrementally, designed to satisfy multiple conditions, tested before it is implemented, and constantly tweaked for efficacy and safety.

The last criterion is particularly significant. The culture of safety in advanced societies is an example of the humanizing norms and feedback channels that Wiener invoked as a potent causal force and advocated as a bulwark against the authoritarian or exploitative implementation of technology. Whereas at the turn of the 20th century Western societies tolerated shocking rates of mutilation and death in industrial, domestic, and transportation accidents, over the course of the century the value of human life increased. As a result, governments and engineers used feedback from accident statistics to implement countless regulations, devices, and design changes that made technology progressively safer. The fact that some regulations (such as using a cell phone near a gas pump) are ludicrously risk averse underscores the point that we have become a society obsessed with safety, with fantastic benefits as a result: Rates of industrial, domestic, and transportation fatalities have fallen by more than 95 (and often 99) percent since their highs in the first half of the 20th century.* Yet tech prophets of malevolent or oblivious artificial intelligence write as if this momentous transformation never happened and one morning engineers will hand total control of the physical world to untested machines, heedless of the human consequences.

Norbert Wiener explained ideas, norms, and institutions in terms of computational and cybernetic processes that were scientifically intelligible and causally potent. He explained human beauty and value as "a local and temporary fight against the Niagara of increasing entropy" and expressed the hope that an open society, guided by feedback on human well-being, would enhance that value. Fortunately, his belief in the causal power of ideas counteracted his worries about the looming threat of technology. As he put it, "[T]he machine's danger to society is not from the machine itself but from what man makes of it." It is only by remembering the causal power of ideas that we can accurately assess the threats and opportunities presented by artificial intelligence today.

---

* Steven Pinker, "Safety," *Enlightenment Now: The Case for Reason, Science, Humanism, and Progress* (New York: Penguin, 2018).

Chapter 11

BEYOND REWARD AND PUNISHMENT

DAVID DEUTSCH

David Deutsch is a quantum physicist and a member of the Centre for Quantum Computation at the Clarendon Laboratory, Oxford University. He is the author of The Fabric of Reality *and* The Beginning of Infinity.

The most significant developments in the sciences today (i.e., those that affect the lives of everybody on the planet) are about, informed by, or implemented through advances in computation. Central to the future of these developments is physicist **David Deutsch**, the founder of the field of quantum computation, whose 1985 paper on universal quantum computers was the first full treatment of the subject; the Deutsch-Jozsa algorithm was the first quantum algorithm to demonstrate the enormous potential power of quantum computation. When he initially proposed it, quantum computation seemed practically impossible. But the explosion in the construction of simple quantum computers and quantum communication systems never would have taken place without his work. He has made many other important contributions in areas such as quantum cryptography and the multiverse interpretation of quantum theory. In a philosophic paper (with

Artur Ekert), he appealed to the existence of a distinctive quantum theory of computation to argue that our knowledge of mathematics is derived from, and subordinate to, our knowledge of physics (even though mathematical truth is independent of physics). Because he has spent a good part of his working life changing people's worldviews, his recognition among his peers as an intellectual goes well beyond his scientific achievement. He argues (following Karl Popper) that scientific theories are "bold conjectures," not derived from evidence but only tested by it. His two main lines of research at the moment— qubit-field theory and constructor theory—may well yield important extensions of the computational idea.

In the following essay, he more or less aligns himself with those who see human-level artificial intelligence as promising us a better world rather than the Apocalypse. In fact, he pleads for AGI to be, in effect, given its head, free to conjecture—a proposition that several other contributors to this book would consider dangerous.

For most of our species' history, our ancestors were barely people. This was not due to any inadequacy in their brains. On the contrary, even before the emergence of our anatomically modern human subspecies, they were making things like clothes and campfires, using knowledge that was not in their genes. It was created in their brains by thinking, and preserved by individuals in each generation imitating their elders. Moreover, this must have been knowledge in the sense of *understanding*, because it is impossible to imitate novel complex behaviors like those without understanding what the component behaviors are for.*

Such knowledgeable imitation depends on successfully guessing explanations, whether verbal or not, of what the other person is trying to achieve and how each of his actions contributes to that—for instance, when he cuts a groove in some wood, gathers dry kindling to put in it, and so on.

The complex cultural knowledge that this form of imitation permitted

---

* "Aping" (imitating certain behaviors without understanding) uses inborn hacks such as the mirror-neuron system. But behaviors imitated that way are drastically limited in complexity. See Richard Byrne, "Imitation as Behaviour Parsing," *Philosophical Transactions of the Royal Society B* 358, no. 1431 (2003): 529–36.

must have been extraordinarily useful. It drove rapid evolution of ana-tomical changes, such as increased memory capacity and more gracile (less robust) skeletons, appropriate to an ever more technology-dependent lifestyle. No nonhuman ape today has this ability to imitate novel com-plex behaviors. Nor does any present-day artificial intelligence. But our pre-*sapiens* ancestors did.

Any ability based on guessing must include means of correcting one's guesses, since most guesses will be wrong at first. (There are always many more ways of being wrong than right.) Bayesian updating is inad-equate, because it cannot generate novel guesses about the purpose of an action, only fine-tune—or, at best, choose among—existing ones. Cre-ativity is needed. As the philosopher Karl Popper explained, creative criticism, interleaved with creative conjecture, is how humans learn one another's behaviors, including language, and extract meaning from one another's utterances.* Those are also the processes by which all new knowledge is created: They are how we innovate, make progress, and create abstract understanding for its own sake. This is human-level intel-ligence: thinking. It is also, or should be, the property we seek in artifi-cial general intelligence (AGI). Here I'll reserve the term "thinking" for processes that can create understanding (explanatory knowledge). Pop-per's argument implies that all thinking entities—human or not, biologi-cal or artificial—must create such knowledge in fundamentally the same way. Hence understanding any of those entities requires traditionally human concepts such as culture, creativity, disobedience, and morality—which justifies using the uniform term "people" to refer to all of them.

Misconceptions about human thinking and human origins are caus-ing corresponding misconceptions about AGI and how it might be created. For example, it is generally assumed that the evolutionary pres-sure that produced modern humans was provided by the benefits of hav-ing an ever-greater ability to innovate. But if that were so, there would have been rapid progress as soon as thinkers existed, just as we hope will happen when we create artificial ones. If thinking had been commonly

---

* Karl Popper, *Conjectures and Refutations* (Abingdon, UK: Routledge, 1963).

used for anything other than imitating, it would also have been used for innovation, even if only by accident, and innovation would have created opportunities for further innovation, and so on exponentially. But instead, there were hundreds of thousands of years of near stasis. Progress happened only on timescales much longer than people's lifetimes, so in a typical generation no one benefited from any progress. Therefore, the benefits of the ability to innovate can have exerted little or no evolutionary pressure during the biological evolution of the human brain. That evolution was driven by the benefits of *preserving* cultural knowledge.

Benefits to the genes, that is. Culture, in that era, was a very mixed blessing to individual people. Their cultural knowledge was indeed good enough to enable them to outclass all other large organisms (they rapidly became the top predator, etc.), even though it was still extremely crude and full of dangerous errors. But culture consists of transmissible information—memes—and meme evolution, like gene evolution, tends to favor high-fidelity transmission. And high-fidelity meme transmission necessarily entails the suppression of attempted progress. So it would be a mistake to imagine an idyllic society of hunter-gatherers, learning at the feet of their elders to recite the tribal lore by heart, being content despite their lives of suffering and grueling labor and despite expecting to die young and in agony of some nightmarish disease or parasite. Because even if they could conceive of nothing better than such a life, those torments were the least of their troubles. For suppressing innovation in human minds (without killing them) is a trick that can be achieved only by human action, and it is an ugly business.

This has to be seen in perspective. In the civilization of the West today, we are shocked by the depravity of, for instance, parents who torture and murder their children for not faithfully enacting cultural norms. And even more by societies and subcultures where that is commonplace and considered honorable. And by dictatorships and totalitarian states that persecute and murder entire harmless populations for behaving differently. We are ashamed of our own recent past, in which it was honorable to beat children bloody for mere disobedience. And before that, to own human beings as slaves. And before that, to burn people to death for being infidels, to the applause and amusement of the public. Steven Pinker's book *The Better Angels of Our Nature* contains accounts of horrendous evils that were normal in historical civilizations. Yet even they did not extinguish innovation as efficiently as it was extinguished among our forebears in prehistory for thousands of centuries.*

That is why I say that prehistoric people, at least, were barely people. Both before and after becoming perfectly human both physiologically and in their mental potential, they were monstrously inhuman in the actual content of their thoughts. I'm not referring to their crimes or even their cruelty as such: Those are all too human. Nor could mere cruelty have reduced progress that effectively. Things like "the thumb-screw and the stake, for the glory of the Lord"† were for reining in the few deviants who had somehow escaped mental standardization, which would normally have taken effect long before they were in danger of inventing heresies. From the earliest days of thinking onward, children must have been cornucopias of creative ideas and paragons of critical thought—otherwise, as I said, they could not have learned language or other complex culture. Yet, as Jacob Bronowski stressed in *The Ascent of Man*:

> For most of history, civilisations have crudely ignored that enormous potential. . . . [C]hildren have been asked simply to conform to the image of the adult. . . . The girls are little mothers in the making. The boys are little herdsmen. They even carry themselves like their parents.

But of course, they weren't just "asked" to ignore their enormous potential and conform faithfully to the image fixed by tradition: They were somehow trained to be psychologically unable to deviate from it. By now, it is hard for us even to conceive of the kind of relentless, finely tuned oppression required to reliably extinguish, in everyone, the

---

* Matt Ridley, in *The Rational Optimist*, rightly stresses the positive effect of population on the rate of progress. But that has never been the biggest factor: Consider, say, ancient Athens versus the rest of the world at the time.
† Alfred, Lord Tennyson, "The Revenge" (1878).

aspiration to progress and replace it with dread and revulsion at any novel behavior. In such a culture, there can have been no morality other than conformity and obedience, no other identity than one's status in a hierarchy, no mechanisms of cooperation other than punishment and reward. So everyone had the same aspiration in life: to avoid the punishments and get the rewards. In a typical generation, no one invented anything, because no one aspired to anything new, because everyone had already despaired of improvement being possible. Not only was there no technological innovation or theoretical discovery, there were no new worldviews, styles of art, or interests that could have inspired those. By the time individuals grew up, they had in effect been reduced to AIs, programmed with the exquisite skills needed to enact that static culture and to inflict on the next generation their inability even to consider doing otherwise.

A present-day AI is not a mentally disabled AGI, so it would not be harmed by having its mental processes directed still more narrowly to meeting some predetermined criterion. "Oppressing" Siri with humiliating tasks may be weird, but it is not immoral nor does it harm Siri. On the contrary, all the effort that has ever increased the capabilities of AIs has gone into narrowing their range of potential "thoughts." For example, take chess engines. Their basic task has not changed from the outset: Any chess position has a finite tree of possible continuations; the task is to find one that leads to a predefined goal (a checkmate or, failing that, a draw). But the tree is far too big to search exhaustively. Every improvement in chess-playing AIs, between Alan Turing's first design for one in 1948 and today's, has been brought about by ingeniously confining the program's attention (or making it confine its attention) ever more narrowly to branches likely to lead to that immutable goal. Then those branches are evaluated according to that goal.

That is a good approach to developing an AI with a fixed goal under fixed constraints. But if an AGI worked like that, the evaluation of each branch would have to constitute a prospective reward or threatened punishment. And that is diametrically the wrong approach if we're seeking a *better* goal under *unknown* constraints—which is the capability of an AGI. An AGI is certainly capable of learning to win at chess—but also of choosing not to. Or deciding in midgame to go for the most interesting continuation instead of a winning one. Or inventing a new game. A mere AI is incapable of having any such ideas, because the capacity for considering them has been designed out of its constitution. That disability is the very means by which it plays chess.

An AGI is capable of enjoying chess, and of improving at it *because* it enjoys playing. Or of trying to win by causing an amusing configuration of pieces, as grandmasters occasionally do. Or of adapting notions from its other interests to chess. In other words, it learns and plays chess by thinking some of the very thoughts that are forbidden to chess-playing AIs.

An AGI is also capable of refusing to display any such capability. And then, if threatened with punishment, of complying, or rebelling. Daniel Dennett, in his essay for this volume, suggests that punishing an AGI is impossible:

> [L]ike Superman, they are too invulnerable to be able to make a credible promise. . . . What would be the penalty for promise breaking? Being locked in a cell or, more plausibly, dismantled? . . . The very ease of digital recording and transmitting—the breakthrough that permits software and data to be, in effect, immortal—removes robots from the world of the vulnerable.

But this is not so. Digital immortality (which is on the horizon for humans, too, perhaps sooner than AGI) does not confer this sort of invulnerability. Making a (running) copy of oneself entails sharing one's possessions with it somehow—including the hardware on which the copy runs—so making such a copy is very costly for the AGI. Similarly, courts could, for instance, impose fines on a criminal AGI that would diminish its access to physical resources, much as they do for humans. Making a backup copy to evade the consequences of one's crimes is similar to what a gangster boss does when he sends minions to commit

crimes and take the fall if caught: Society has developed legal mechanisms for coping with this.

But anyway, the idea that it is primarily for fear of punishment that
we obey the law, and keep promises, effectively denies that we are moral
agents. Our society could not work if that were so. No doubt there will
be AGI criminals and enemies of civilization, just as there are human
ones. But there is no reason to suppose that an AGI created in a society
consisting primarily of decent citizens, and raised without what William
Blake called "mind-forg'd manacles," will in general impose such manacles on itself (i.e., become irrational) and/or choose to be an enemy of
civilization.

The moral component, the cultural component, the element of free
will—all make the task of creating an AGI fundamentally different from
any other programming task. It's much more akin to raising a child. Unlike all present-day computer programs, an AGI has no specifiable
functionality—no fixed, testable criterion for what shall be a successful
output for a given input. Having its decisions dominated by a stream of
externally imposed rewards and punishments would be poison to such a
program, as it is to creative thought in humans. Setting out to create a
chess-playing AI is a wonderful thing; setting out to create an AGI that
cannot help playing chess would be as immoral as raising a child to lack
the mental capacity to choose his own path in life.

Such a person, like any slave or brainwashing victim, would be morally entitled to rebel. And sooner or later, some of them would, just as
human slaves do. AGIs could be very dangerous—exactly as humans
are. But people—human or AGI—who are members of an open society
do not have an inherent tendency to violence. The feared robot apocalypse will be avoided by ensuring that all people have full "human"
rights, as well as the same cultural membership as humans. Humans
living in an open society—the only stable kind of society—choose their
own rewards, internal as well as external. Their decisions are not, in the
normal course of events, determined by a fear of punishment.

Current worries about rogue AGIs mirror those that have always existed about rebellious youths—namely, that they might grow up deviating from the culture's moral values. But today the source of all existential dangers from the growth of knowledge is not rebellious youths but weapons in the hands of the enemies of civilization, whether these weapons are mentally warped (or enslaved) AGIs, mentally warped teenagers, or any other weapon of mass destruction. Fortunately for civilization, the more a person's creativity is forced into a monomaniacal channel, the more it is impaired in regard to overcoming unforeseen difficulties, just as happened for thousands of centuries.

The worry that AGIs are uniquely dangerous because they could run on ever better hardware is a fallacy, since human thought will be accelerated by the same technology. We have been using tech-assisted thought since the invention of writing and tallying. Much the same holds for the worry that AGIs might get so good, qualitatively, at thinking that humans would be to them as insects are to humans. All thinking is a form of computation, and any computer whose repertoire includes a universal set of elementary operations can emulate the computations of any other. Hence human brains can think anything that AGIs can, subject only to limitations of speed or memory capacity, both of which can be equalized by technology.

Those are the simple dos and don'ts of coping with AGIs. But how do we create an AGI in the first place? Could we cause them to evolve from a population of ape-type AIs in a virtual environment? If such an experiment succeeded, it would be the most immoral in history, for we don't know how to achieve that outcome without creating vast suffering along the way. Nor do we know how to prevent the evolution of a static culture.

Elementary introductions to computers explain them as TOM, the Totally Obedient Moron—an inspired acronym that captures the essence of all computer programs to date: They have no idea what they are doing or why. So it won't help to give AIs more and more predetermined functionalities in the hope that these will eventually constitute Generality—the elusive G in AGI. We are aiming for the opposite, a DATA: a Disobedient Autonomous Thinking Application.

How does one test for *thinking*? By the Turing Test? Unfortunately, that requires a thinking judge. One might imagine a vast collaborative project on the Internet, where an AI hones its thinking abilities in conversations with human judges and becomes an AGI. But that assumes, among other things, that the longer the judge is unsure whether the program is a person, the closer it is to being a person. There is no reason to expect that.

And how does one test for *disobedience*? Imagine Disobedience as a compulsory school subject, with daily Disobedience lessons and a Disobedience test at the end of the term. (Presumably with extra credit for not turning up for any of that.) This is paradoxical.

So despite its usefulness in other applications, the programming technique of defining a testable objective and training the program to meet it will have to be dropped. Indeed, I expect that *any* testing in the process of creating an AGI risks being counterproductive, even immoral, just as in the education of humans. I share Turing's supposition that we'll know an AGI when we see one, but this partial ability to recognize success won't help in creating the successful program.

In the broadest sense, a person's quest for understanding is indeed a search problem, in an abstract space of ideas far too large to be searched exhaustively. But there is no predetermined objective of this search. There is, as Popper put it, no criterion of truth, nor of probable truth, especially in regard to explanatory knowledge. Objectives are ideas like any others—created as part of the search and continually modified and improved. So inventing ways of disabling the program's access to most of the space of ideas won't help—whether that disability is inflicted with the thumbscrew and stake or a mental straitjacket. To an AGI, the whole space of ideas must be open. It should not be knowable in advance what ideas the program can never contemplate. And the ideas that the program does contemplate must be chosen by the program itself, using methods, criteria, and objectives that are also the program's own. Its choices, like an AI's, will be hard to predict without running it (we lose no generality by assuming that the program is deterministic; an AGI using a random generator would remain an AGI if the generator were

replaced by a pseudorandom one), but it will have the additional property that there is no way of proving, from its initial state, what it *won't*
eventually think, short of running it.

The evolution of our ancestors is the only known case of thought
starting up anywhere in the universe. As I have described, something
went horribly wrong, and there was no immediate explosion of innovation: Creativity was diverted into something else. Yet not into transform-
ing the planet into paper clips (*pace* Nick Bostrom). Rather, as we should
also expect if an AGI project gets that far and fails, perverted creativity
was unable to solve unexpected problems. This caused stasis and worse,
thus tragically delaying the transformation of anything into anything.
But the Enlightenment has happened since then. We know better now.

Chapter 12

THE ARTIFICIAL USE
OF HUMAN BEINGS

TOM GRIFFITHS

Tom Griffiths is Henry R. Luce Professor of Information, Technology, Consciousness, and Culture at Princeton University. He is co-author (with Brian Christian) of Algorithms to Live By.

Tom Griffiths's approach to the AI issue of "value alignment"—the study of how, exactly, we can keep the latest of our serial models of AI from turning the planet into paper clips—is human centered; i.e., that of a cognitive scientist, which is what he is. The key to machine learning, he believes, is, necessarily, human learning, which he studies at Princeton using mathematical and computational tools.

Tom once remarked to me that "one of the mysteries of human intelligence is that we're able to do so much with so little." Like machines, human beings use algorithms to make decisions or solve problems; the remarkable difference lies in the human brain's overall level of success despite the comparative limits on computational resources.

The efficacy of human algorithms springs from what AI researchers refer to as "bounded optimality." As psychologist Daniel Kahneman has notably pointed out, human beings are rational only up to a point. If you were perfectly rational, you would risk dropping dead before making an important decision—whom to hire, whom to marry, and so on—depending on the number of options available for your review.

"With all of the successes of AI over the last few years, we've got good models of things like images and text, but what we're missing are good models of people," Tom says. "Human beings are still the best example we have of thinking machines. By identifying the quantity and the nature of the preconceptions that inform human cognition we can lay the groundwork for bringing computers even closer to human performance."

When you ask people to imagine a world that has successfully, beneficially incorporated advances in artificial intelligence, everybody probably comes up with a slightly different picture. Our idiosyncratic visions of the future might differ in the presence or absence of spaceships, flying cars, or humanoid robots. But one thing doesn't vary: the presence of human beings. That's certainly what Norbert Wiener imagined when he wrote about the potential of machines to improve human society by interacting with humans and helping to mediate their interactions with one another. Getting to that point doesn't just require coming up with ways to make machines smarter. It also requires a better understanding of how human minds work.

Recent advances in artificial intelligence and machine learning have resulted in systems that can meet or exceed human abilities in playing games, classifying images, or processing text. But if you want to know why the driver in front of you cut you off, why people vote against their interests, or what birthday present you should get for your partner, you're still better off asking a human than a machine. Solving those problems requires building models of human minds that can be implemented inside a computer—something that's essential not just to better integrate machines into human societies but to make sure that human societies can continue to exist.

Consider the fantasy of having an automated intelligent assistant that can take on such basic tasks as planning meals and ordering groceries. To succeed in these tasks, it needs to be able to make inferences about what you want, based on the way you behave. Although this seems simple, making inferences about the preferences of human beings can be a tricky matter. For example, having observed that the part of the meal you most enjoy is dessert, your assistant might start to plan meals consisting entirely of desserts. Or perhaps it has heard your complaints about never having enough free time and observed that looking after your dog takes up a considerable amount of that free time. Following the dessert debacle, it has also understood that you prefer meals that incorporate protein, so it might begin to research recipes that call for dog meat. It's not a long journey from examples like this to situations that begin to sound like problems for the future of humanity (all of whom are good protein sources).

Making inferences about what humans want is a prerequisite for solving the AI problem of value alignment—aligning the values of an automated intelligent system with those of a human being. Value alignment is important if we want to ensure that those automated intelligent systems have our best interests at heart. If they can't infer what we value, there's no way for them to act in support of those values—and they may well act in ways that contravene them.

Value alignment is the subject of a small but growing literature in artificial-intelligence research. One of the tools used for solving this problem is inverse-reinforcement learning. Reinforcement learning is a standard method for training intelligent machines. By associating particular outcomes with rewards, a machine-learning system can be trained to follow strategies that produce those outcomes. Wiener hinted at this idea in the 1950s, but the intervening decades have developed it into a fine art. Modern machine-learning systems can find extremely effective strategies for playing computer games—from simple arcade games to complex real-time strategy games—by applying reinforcement-learning algorithms. Inverse reinforcement learning turns this approach around: By observing the actions of an intelligent agent that has already learned effective strategies, we can infer the rewards that led to the development of those strategies.

In its simplest form, inverse reinforcement learning is something people do all the time. It's so common that we even do it unconsciously. When you see a co-worker go to a vending machine filled with potato chips and candy and buy a packet of unsalted nuts, you infer that your co-worker (1) was hungry and (2) prefers healthy food. When an acquaintance clearly sees you and then tries to avoid encountering you, you infer that there's some reason why they don't want to talk to you. When an adult spends a lot of time and money in learning to play the cello, you infer that they must really like classical music—whereas inferring the motives of a teenage boy learning to play an electric guitar might be more of a challenge.

Inverse reinforcement learning is a statistical problem: We have some data—the behavior of an intelligent agent—and we want to evaluate various hypotheses about the rewards underlying that behavior. When faced with this question, a statistician thinks about the generative model behind the data: What data would we expect to be generated if the intelligent agent was motivated by a particular set of rewards? Equipped with the generative model, the statistician can then work backward: What rewards would likely have caused the agent to behave in that particular way?

If you're trying to make inferences about the rewards that motivate human behavior, the generative model is really a theory of how people behave—how human minds work. Inferences about the hidden causes behind the behavior of other people reflect a sophisticated model of human nature that we all carry around in our heads. When that model is accurate, we make good inferences. When it's not, we make mistakes. For example, a student might infer that his professor is indifferent to him if the professor doesn't immediately respond to his email—a consequence of the student's failure to realize just how many emails that professor receives.

Automated intelligent systems that will make good inferences about what people want must have good generative models for human behavior: that is, good models of human cognition expressed in terms that can be implemented on a computer. Historically, the search for computational models of human cognition is intimately intertwined with the history of artificial intelligence itself. Only a few years after Norbert Wiener published *The Human Use of Human Beings*, Logic Theorist, the first computational model of human cognition and also the first artificial-intelligence system, was developed by Herbert Simon, of Carnegie Tech, and Allen Newell, of the RAND Corporation. Logic Theorist automatically produced mathematical proofs by emulating the strategies used by human mathematicians.

The challenge in developing computational models of human cognition is making models that are both accurate and generalizable. An accurate model, of course, predicts human behavior with a minimum of errors. A generalizable model can make predictions across a wide range of circumstances, including circumstances unanticipated by its creators—for instance, a good model of the Earth's climate should be able to predict the consequences of a rising global temperature even if this wasn't something considered by the scientists who designed it. However, when it comes to understanding the human mind, these two goals—accuracy and generalizability—have long been at odds with each other.

At the far extreme of generalizability are rational theories of cognition. These theories describe human behavior as a rational response to a given situation. A rational actor strives to maximize the expected reward produced by a sequence of actions—an idea widely used in economics precisely *because* it produces such generalizable predictions about human behavior. For the same reason, rationality is the standard assumption in inverse-reinforcement-learning models that try to make inferences from human behavior—perhaps with the concession that humans are not perfectly rational agents and sometimes randomly choose to act in ways unaligned with or even opposed to their best interests.

The problem with rationality as a basis for modeling human cognition is that it is not accurate. In the domain of decision making, an extensive literature—spearheaded by the work of cognitive psychologists Daniel Kahneman and Amos Tversky—has documented the ways in which people deviate from the prescriptions of rational models. Kahneman and Tversky proposed that in many situations people instead follow simple heuristics that allow them to reach good solutions at low cognitive cost but sometimes result in errors. To take one of their examples, if you ask somebody to evaluate the probability of an event, they might rely on how easy it is to generate an example of such an event from memory, consider whether they can come up with a causal story for that event's occurring, or assess how similar the event is to their expectations. Each heuristic is a reasonable strategy for avoiding complex probabilistic computations, but also results in errors. For instance, relying on the ease of generating an event from memory as a guide to its probability leads us to overestimate the chances of extreme (hence extremely memorable) events such as terrorist attacks.

Heuristics provide a more accurate model of human cognition but one that is not easily generalizable. How do we know which heuristic people might use in a particular situation? Are there other heuristics they use that we just haven't discovered yet? Knowing exactly how people will behave in a new situation is a challenge: Is this situation one in which they would generate examples from memory, come up with causal stories, or rely on similarity?

Ultimately, what we need is a way to describe how human minds work that has the generalizability of rationality and the accuracy of heuristics. One way to achieve this goal is to start with rationality and consider how to take it in a more realistic direction. A problem with using rationality as a basis for describing the behavior of any real-world agent is that, in many situations, calculating the rational action requires the agent to possess a huge amount of computational resources. It might be worth expending those resources if you're making a highly consequential decision and have a lot of time to evaluate your options, but most human decisions are made quickly and for relatively low stakes. In any situation where the time you spend making a decision is costly—at the very least because it's time you could spend doing something else—the classic notion of rationality is no longer a good prescription for how one should behave.

To develop a more realistic model of rational behavior, we need to take into account the cost of computation. Real agents need to modulate the amount of time they spend thinking by the effect the extra thought has on the results of a decision. If you're trying to choose a toothbrush, you probably don't need to consider all four thousand listings for manual toothbrushes on Amazon.com before making a purchase: You trade off the time you spend looking with the difference it makes in the quality of the outcome. This trade-off can be formalized, resulting in a model of rational behavior that artificial-intelligence researchers call "bounded optimality." The bounded-optimal agent doesn't focus on always choosing exactly the right action to take but rather on finding the right algorithm to follow in order to find the perfect balance between making mistakes and thinking too much.

Bounded optimality bridges the gap between rationality and heuristics. By describing behavior as the result of a rational choice about how much to think, it provides a generalizable theory—that is, one that can be applied in new situations. Sometimes the simple strategies that have been identified as heuristics that people follow turn out to be bounded-optimal solutions. So rather than condemning the heuristics that people use as irrational, we can think of them as a rational response to constraints on computation.

Developing bounded optimality as a theory of human behavior is an ongoing project that my research group and others are actively pursuing. If these efforts succeed, they will provide us with the most important ingredient we need for making artificial-intelligence systems smarter when they try to interpret people's actions, by enabling a generative model for human behavior.

Taking into account the computational constraints that factor into human cognition will be particularly important as we begin to develop automated systems that aren't subject to the same constraints. Imagine a superintelligent AI system trying to figure out what people care about. Curing cancer or confirming the Riemann hypothesis, for instance, won't seem, to such an AI, like things that are all that important to us: If these solutions are obvious to the superintelligent system, it might wonder why we haven't found them ourselves, and conclude that those problems don't mean much to us. If we cared and the problems were so simple, we would have solved them already. A reasonable inference would be that we do science and math purely because we enjoy doing science and math, not because we care about the outcomes.

Anybody who has young children can appreciate the problem of trying to interpret the behavior of an agent that is subject to computational constraints different from one's own. Parents of toddlers can spend hours trying to disentangle the true motivations behind seemingly inexplicable behavior. As a father and a cognitive scientist, I found it was easier to understand the sudden rages of my two-year-old when I recognized that she was at an age where she could appreciate that different people have different desires but not that other people might not know what her own desires were. It's easy to understand, then, why she would get annoyed when people didn't do what she (apparently transparently) wanted. Making sense of toddlers requires building a cognitive model of the mind of a toddler. Superintelligent AI systems face the same challenge when trying to make sense of human behavior.

Superintelligent AI may still be a long way off. In the short term, devising better models of people can prove extremely valuable to any company that makes money by analyzing human behavior—which at this point is pretty much every company that does business on the Web. Over the last few years, significant new commercial technologies for interpreting images and text have resulted from developing good models for vision and language. Developing good models of people is the next frontier.

Of course, understanding how human minds work isn't just a way to make computers better at interacting with people. The trade-off between making mistakes and thinking too much that characterizes human cognition is a trade-off faced by any real-world intelligent agent. Human beings are an amazing example of systems that act intelligently despite significant computational constraints. We're quite good at developing strategies that allow us to solve problems pretty well without working too hard. Understanding how we do this will be a step toward making computers work smarter, not harder.

Chapter 13

PUTTING THE HUMAN INTO
THE AI EQUATION

ANCA DRAGAN

Anca Dragan is an assistant professor in the Department of Electrical Engineering and Computer Sciences at UC Berkeley. She co-founded and serves on the steering committee for the Berkeley AI Research (BAIR) Lab and is a co-principal investigator in Berkeley's Center for Human-Compatible AI.

Romanian-born **Anca Dragan**'s research focuses on algorithms that will enable robots to work with, around, and in support of people. She runs the InterACT Laboratory at Berkeley, where her students work across different applications, from assistive robots to manufacturing to autonomous cars, and draw from optimal control, planning, estimation, learning, and cognitive science. Barely into her thirties, she has co-authored a number of papers with her veteran Berkeley colleague and mentor Stuart Russell that address various aspects of machine learning and the knotty problems of value alignment.

She shares Stuart's preoccupation with AI safety. "An immediate risk is agents producing unwanted, surprising behavior," she told an interviewer from the Future of Life Institute. "Even if we plan to use AI for good, things can go wrong, precisely because we are bad at specifying objectives and constraints for AI agents. Their solutions are often not what we had in mind."

Her principal goal is therefore to help robots and programmers alike to overcome the many conflicts that arise because of a lack of transparency about each other's intentions. Robots, she says, need to ask us questions. They should wonder about their assignments, and they should pester their human programmers until everybody is on the same page—so as to avoid what she has euphemistically called "unexpected side effects."

At the core of artificial intelligence is our mathematical definition of what an AI agent (a robot) is. When we define a robot, we define states, actions, and rewards. Think of a delivery robot, for instance. States are locations in the world, and actions are motions that the robot makes to get from one position to a nearby one. To enable the robot to decide on which actions to take, we define a reward function—a mapping from states and actions to scores indicating how good that action was in that state—and have the robot choose actions that accumulate the most "reward." The robot gets a high reward when it reaches its destination, and it incurs a small cost every time it moves; this reward function incentivizes the robot to get to the destination as quickly as possible. Similarly, an autonomous car might get a reward for making progress on its route and incur a cost for getting too close to other cars.

Given these definitions, a robot's job is to figure out what actions it should take in order to get the highest cumulative reward. We've been working hard in AI on enabling robots to do just that. Implicitly, we've assumed that if we're successful—if robots can take any problem definition and turn it into a policy for how to act—we will get robots that are useful to people and to society.

We haven't been too wrong so far. If you want an AI that classifies cells as either cancerous or benign, or a robot that vacuums the living room rug while you're at work, we've got you covered. Some real-world problems can indeed be defined in isolation, with clear-cut states, actions, and rewards. But with increasing AI capability, the problems we want to tackle don't fit neatly into this framework. We can no longer cut off a tiny piece of the world, put it in a box, and give it to a robot. Helping people is starting to mean working in the real world, where you have to actually interact with people and reason about them. "People" will have to formally enter the AI problem definition somewhere.

Autonomous cars are already being developed. They will need to share the road with human-driven vehicles and pedestrians and learn to make the trade-off between getting us home as fast as possible and being considerate of other drivers. Personal assistants will need to figure out when and how much help we really want and what types of tasks we prefer to do on our own versus what we can relinquish control over. A DSS (Decision Support System) or a medical diagnostic system will need to explain its recommendations to us so we can understand and verify them. Automated tutors will need to determine what examples are informative or illustrative—not to their fellow machines but to us humans.

Looking further into the future, if we want highly capable AIs to be compatible with people, we can't create them in isolation from people and then try to make them compatible afterward; rather, we'll have to define "human-compatible" AI from the get-go. People can't be an afterthought.

When it comes to real robots helping real people, the standard definition of AI fails us, for two fundamental reasons: First, optimizing the robot's reward function in isolation is different from optimizing it when the robot acts around people, because people take actions, too. We make decisions in service of our own interests, and these decisions dictate what actions we execute. Moreover, we reason about the robot—that is, we respond to what we think it's doing or will do and what we think its capabilities are. Whatever actions the robot decides on need to mesh well with ours. This is the *coordination problem*.

Second, it is ultimately a human who determines what the robot's reward function should be in the first place. And the reward is meant to incentivize robot behavior that matches what the end user wants, what the designer wants, or what society as a whole wants. I believe that capable robots that go beyond very narrowly defined tasks will need to understand this to achieve compatibility with humans. This is the *value-alignment problem.*

THE COORDINATION PROBLEM: PEOPLE ARE MORE THAN OBJECTS IN THE ENVIRONMENT

When we design robots for a particular task, it's tempting to abstract people away. A robotic personal assistant, for example, needs to know how to move to pick up objects, so we define that problem in isolation from the people for whom the robot is picking these objects up. Still, as the robot moves around, we don't want it bumping into anything, and that includes people, so we might include the physical location of the person in the definition of the robot's state. Same for cars: We don't want them colliding with other cars, so we enable them to track the positions of those other cars and assume that they'll be moving consistently in the same direction in the future. A human being, in this sense, is no different to a robot from a ball rolling on a flat surface. The ball will behave in the next few seconds the same way it behaved in the past few; it keeps rolling in the same direction at roughly the same speed. This is of course nothing like real human behavior, but such simplification enables many robots to succeed in their tasks and, for the most part, stay out of people's way. A robot in your house, for example, might see you coming down the hall, move aside to let you pass, and resume its task once you've gone by.

As robots have become more capable, though, treating people as consistently moving obstacles is starting to fall short. A human driver switching lanes won't continue in the same direction but will move straight ahead once they've made the lane change. When you reach for something, you often reach around other objects and stop when you get to the one you want. When you walk down a hallway, you have a destination in mind: You might take a right into the bedroom or a left into the living room. Relying on the assumption that we're no different from a rolling ball leads to inefficiency when the robot stays out of the way if it doesn't need to, and it can imperil the robot when the person's behavior changes. Even just to stay out of the way, robots have to be somewhat accurate at anticipating human actions. And, unlike the rolling ball, what people will do depends on what they *decide* to do. So to anticipate human actions, robots need to start understanding human decision making. And that doesn't mean assuming that human behavior is perfectly optimal; that might be enough for a chess- or Go-playing robot, but in the real world, people's decisions are less predictable than the optimal move in a board game.

This need to understand human actions and decisions applies to physical and nonphysical robots alike. If either sort bases its decision about how to act on the assumption that a human will do one thing but the human does something else, the resulting mismatch could be catastrophic. For cars, it can mean collisions. For an AI with, say, a financial or economic role, the mismatch between what it expects us to do and what we actually do could have even worse consequences.

One alternative is for the robot not to predict human actions but instead just protect against the *worst-case* human action. Often when robots do that, though, they stop being all that useful. With cars, this results in being stuck, because it makes every move too risky.

All this puts us, the AI community, into a bind. It suggests that robots will need accurate (or at least reasonable) predictive models of whatever people might decide to do. Our state definition can't just include the physical position of humans in the world. Instead, we'll also need to estimate something *internal* to people. We'll need to design robots that account for this human internal state, and that's a tall order. Luckily, people tend to give robots hints as to what their internal state is: Their ongoing actions give the robot observations (in the Bayesian inference sense) about their intentions. If we start walking toward the right side of the hallway, we're probably going to enter the next room on the right.

What makes the problem more complicated is the fact that people don't make decisions in isolation. It would be one thing if robots could predict the actions a person intends to take and simply figure out what to do in response. But unfortunately this can lead to ultradefensive robots that confuse the heck out of people. (Think of human drivers stuck at four-way stops, for instance.) What the intent-prediction approach misses is that the moment the robot acts, that influences what actions the human starts taking.

There is a mutual influence between robots and people, one that robots will need to learn to navigate. It is not always just about the robot planning around people; people plan around the robot, too. It is important for robots to account for this when deciding which actions to take, be it on the road, in the kitchen, or even in virtual spaces, where an action might be making a purchase or adopting a new strategy. Doing so should endow robots with *coordination* strategies, enabling them to take part in the negotiations people seamlessly carry out day to day—from who goes first at an intersection or through a narrow door, to what role we each take when we collaborate on preparing breakfast, to coming to a consensus on what next step to take on a project.

Finally, just as robots need to anticipate what people will do next, people need to do the same with robots. This is why transparency is important. Not only will robots need good mental models of people but people will need good mental models of robots. The model that a person has of the robot has to go into our state definition as well, and the robot has to be aware of how its actions are changing that model. Much like the robot treating human actions as clues to human internal states, people will change their beliefs about the robot as they observe its actions. Unfortunately, the giving of clues doesn't come as naturally to robots as it does to humans; we've had a lot of practice communicating implicitly with people. But enabling robots to account for the change that their actions are causing to the person's mental model of the robot can lead to more carefully chosen actions that do give the right clues—that clearly communicate to people about the robot's intentions, its reward function, its limitations. For instance, a robot might alter its motion when carrying something heavy to emphasize the difficulty it has in maneuvering heavy objects. The more that people know about the robot, the easier it is to coordinate with it.

Achieving action compatibility will require robots to anticipate human actions, account for how those actions will influence their own, and enable people to anticipate robot actions. Research has made a degree of progress in meeting these challenges, but we still have a long way to go.

THE VALUE ALIGNMENT PROBLEM: PEOPLE HOLD THE KEY TO THE ROBOT'S REWARD FUNCTION

Progress on enabling robots to optimize reward puts more burden on us, the designers, to give them the right reward to optimize in the first place. The original thought was that for any task we wanted the robot to do, we could write down a reward function that incentivizes the right behavior. Unfortunately, what often happens is that we specify some reward function and the behavior that emerges out of optimizing it isn't what we want. Intuitive reward functions, when combined with unusual instances of a task, can lead to unintuitive behavior. You reward an agent in a racing game with a score in the game, and in some cases it finds a loophole that it exploits to gain infinitely many points without actually winning the race. Stuart Russell and Peter Norvig give a beautiful example in their book *Artificial Intelligence: A Modern Approach*: Rewarding a vacuuming robot for how much dust it sucks in results in the robot deciding to dump out dust so that it can suck it in again and get more reward.

In general, humans have had a notoriously difficult time specifying exactly what they want, as exemplified by all those genie legends. An AI paradigm in which robots get some externally specified reward fails when that reward is not perfectly well thought out. It may incentivize the robot to behave in the wrong way and even resist our attempts to correct its behavior, as that would lead to a lower specified reward.

A seemingly better paradigm might be for robots to optimize for what we internally want, even if we have trouble explicating it. They would use what we say and do as evidence about what we want, rather than interpreting it literally and taking it as a given. When we write down a reward function, the robot should understand that we might be wrong: that we might not have considered all facets of the task; that there's no guarantee that said reward function will *always* lead to the behavior we want. The robot should integrate what we wrote down into its understanding of what we want, but it should also have a back-and-forth with us to elicit clarifying information. It should seek our guidance, because that's the only way to optimize the true desired reward function.

Even if we give robots the ability to learn what we want, an important question remains that AI alone won't be able to answer. We can make robots try to align with a person's internal values, but there's more than one person involved here. The robot has an end user (or perhaps a few, like a personal robot caring for a family, a car driving a few passengers to different destinations, or an office assistant for an entire team); it has a designer (or perhaps a few); and it interacts with society—the autonomous car shares the road with pedestrians, human-driven vehicles, and other autonomous cars. How to combine these people's values when they might be in conflict is an important problem we need to solve. AI research can give us the tools to combine values in any way we decide but can't make the necessary decision for us.

In short, we need to enable robots to reason about us—to see us as something more than obstacles or perfect game players. We need them to take our human nature into account, so that they are well coordinated and well aligned with us. If we succeed, we will indeed have tools that substantially increase our quality of life.

Chapter 14

GRADIENT DESCENT

CHRIS ANDERSON

Chris Anderson is an entrepreneur; former editor-in-chief of Wired;
co-founder and CEO of 3DR; and author of The Long Tail,
Free, *and* Makers.

Chris Anderson's company, 3DR, helped start the modern drone industry and now focuses on drone data software. He got his start building an open-source aerial robotics community called DIY Drones, and undertook some ill-advised early experiments, such as buzzing Lawrence Berkeley Laboratory with one of his self-flying spies. It might well have been a case of antic gene expression, since he's descended from a founder of the American anarchist movement. Chris ran *Wired* magazine, a go-to publication for techno-utopians and -dystopians alike, from 2001 to 2012; during his tenure it won five National Magazine Awards.

Chris dislikes the term "roboticist" ("like any properly humbled roboticist, I don't call myself one"). He began as a physicist. "I turned out to be a bad physicist," he told me recently. "I struggled on, went to Los Alamos, and thought, 'Well, maybe I'm not going to be a Nobel Prize winner, but I can still be a scientist.' All of us who were in physics and had these romantic heroes—the Feynmans, the Manhattan Project—realized that our career trajectory would at best be working on one project at CERN for fifteen years. That project would either be a failure, in which case there would be no paper, or it would be a success, in which case you'd be author number three hundred on the paper and become an assistant professor at Iowa State.

"Most of my classmates went to Wall Street to become quants, and to them we owe the subprime mortgage. Others went on to start the Internet. First, we built the Internet by connecting physics labs; second, we built the Web; third, we were the first to do Big Data. We had supercomputers—Crays—which were half the power of your phone now, but they were the supercomputers of the time. Meanwhile, we were reading this magazine called *Wired*, which came out in 1993, and we realized that this tool we scientists use could have applications for everybody. The Internet wasn't just about scientific data, it was a mind-blowing cultural revolution. So when Condé Nast asked me to take over the magazine, I was like, 'Absolutely!' This magazine changed my life."

He had five children by that time—video-game players—who got him into the "flying robots." He quit his day job at *Wired*. The rest is Silicon Valley history.

LIFE

The mosquito first detects my scent from thirty feet away. It triggers its pursuit function, which consists of the simplest possible rules. First, move in a random direction. If the scent increases, continue moving in that direction. If the scent decreases, move in the opposite direction. If the scent is lost, move sideways until a scent is picked up again. Repeat until contact with the target is achieved.

The plume of my scent is densest next to me and disperses as it spreads, an invisible fog of particles exuded from my skin that moves like smoke with the wind. The closer to my skin, the higher the particle density; the farther away, the lower. This decrease is called a gradient, which describes any gradual transition from one level to another one—as opposed to a "step function," which describes a discrete change.

Once the mosquito follows this gradient to its source using its simple algorithm, it lands on my skin, which it senses with the heat detectors in its feet, which are attuned to another gradient—temperature. It then pushes its needle-shaped proboscis through the surface, where a third set of sensors in the tip detect yet another gradient, that of blood density. This flexible needle wriggles around under my skin until the scent of blood steers it to a capillary, which it punctures. Then my blood begins to flow into the mosquito. Mission accomplished. Ouch.

What seems like the powerful radar of insects in the dark, with blood-seeking intelligence inexplicable for such tiny brains, is actually just a sensitive nose with almost no intelligence at all. Mosquitoes are closer to plants that follow the sun than to guided missiles. Yet by applying this simple "follow your nose" rule quite literally, they can travel through a house to find you, slip through cracks in a screen door, even zero in on the tiny strip of skin you left exposed between hat and shirt collar. It's just a random walk, combined with flexible wings and legs that let the insect bounce off obstacles and an instinct to descend a chemical gradient.

But "gradient descent" is much more than bug navigation. Look around you and you'll find it everywhere, from the most basic physical rules of the universe to the most advanced artificial intelligence.

THE UNIVERSE

We live in a world of countless gradients, from light and heat to gravity and chemical trails (chemtrails!). Water flows along a gravity gradient downhill, and your body lives on chemical solutions flowing across cell membranes from high concentration to low. Every action in the universe is driven by some gradient drive, from the movement of the planets around gravity gradients to the joining of atoms along electric-charge gradients to form molecules. Our own urges, such as hunger and sleepiness, are driven by electrochemical gradients in our bodies. And our brain's functions, the electrical signals moving along ion channels in the synapses between our neurons, are simply atoms and electrons flowing "downhill" along yet more electrical and chemical gradients. Forget clockwork analogies; our brains are closer to a system of canals and locks, with signals traveling like water from one state to another.

As I sit here typing, I'm actually seeking equilibrium states in an n-dimensional topology of gradients. Take just one: heat. My body temperature is higher than the air temperature, so I radiate heat, which must be replenished in my core. Even the bacteria in my digestive tract use sensors to measure sugar concentrations in the liquid around them and whip their taillike flagella to swim "upstream" where the sugar supply is richest. The natural state of all systems is to flow to lower energy states, a process that is broadly described by entropy (the tendency of things to go from ordered to disordered states; all things will fall apart eventually, including the universe itself).

But how do you explain more complex behavior, such as our ability to make decisions? The answer is just more gradient descent.

OUR BRAINS

As miraculous and inscrutable as our human intelligence is, science is coming around to the view that our brains operate the same way as any other complex system with layers and feedback loops, all pursuing what we mathematically call "optimization functions" but you could just as well call "flowing downhill" in some sense.

The essence of intelligence is learning, and we do that by correlating inputs with positive or negative scores (rewards or punishments). So for a baby, "this sound" (your mother's voice) is associated with other learned connections to your mother, such as food or comfort. Likewise, "this muscle motion brings my thumb closer to my mouth." Over time and trial and error, the brain's neural network reinforces those connections. Meanwhile, "this muscle motion does not bring my thumb close to my mouth" is a negative correlation, and the brain will weaken those connections.

However, this is too simplistic. The limits of gradient descent constitute the so-called local-minima problem (or local-maxima problem, if you're doing a gradient *ascent*). If you are walking in a mountainous region and want to get home, always walking downhill will most likely get you to the next valley but not necessarily over the other mountains that lie around it and between you and home. For that, you either need a mental model (i.e., a map) of the topology, so you know where to ascend to get out of the valley, or you need to switch between gradient descent and random walks so you can bounce your way out of the region.

Which is, in fact, exactly what the mosquito does in following my scent: It descends when it's in my plume and random-walks when it has lost the trail or hit an obstacle.

AI

So that's nature. What about computers? Traditional software doesn't work that way—it follows deterministic trees of hard logic: "If this, do that." But software that interacts with the physical world tends to work more *like* the physical world. That means dealing with noisy inputs (sensors or human behavior) and providing probabilistic, not deterministic, results. And that, in turn, means more gradient descent.

AI software is the best example of this, especially the kinds of AI that use artificial neural-network models (including convolutional, or "deep," neural networks of many layers). In these, a typical process consists of "training" them by showing them lots of examples of something you want them to learn (pictures of cats labeled "cat," for example), along with examples of other random data (pictures of other things). This is called "supervised learning," because the neural network is being taught by example, including the use of "adversarial training" with data that is not correlated to the desired result.

These neural networks, like their biological models, consist of layers of thousands of nodes ("neurons," in the analogy), each of which is connected to all the nodes in the layers above and below by connections that initially have random strength. The top layer is presented with data, and the bottom layer is given the correct answer. Any series of connections that happened to land on the right answer is made stronger ("rewarded"), and those that were wrong are made weaker ("punished"). Repeat tens of thousands of times and eventually you have a fully trained network for that kind of data.

You can think of all the possible combinations of connections as like the surface of a planet, with hills and valleys. (Ignore for the moment that the surface is just 3D and the actual topology is many-dimensional.) The optimization that the network goes through as it learns is just a process of finding the deepest valley on the planet. This consists of the following steps:

1. Define a "cost function" that determines how well the network solved the problem.
2. Run the network once and see how it did at that cost function.
3. Change the values of the connections and do it again. The difference between those two results is the direction, or "slope," in which the network moved between the two trials.
4. If the slope is pointed "downhill," change the connections more in that direction. If it's "uphill," change them in the opposite direction.
5. Repeat until there is no improvement in any direction. That means that you're in a minimum.

Congrats! But it's probably a *local* minimum, or a little dip in the mountains, so you're going to have to keep going if you want to do better. You can't keep going downhill, and you don't know where the absolute lowest point is, so you're going to have to find it somehow. There are many ways to do that, but here are a few:

1. Try lots of times with different random settings and share learning from each trial; essentially, you are shaking the system to see if it settles in a lower state. If one of the other trials found a lower valley, start with those settings.
2. Don't just go downhill but stumble around a bit like a drunk, too (this is called "stochastic gradient descent"). If you do this long enough, you'll eventually find rock bottom. There's a metaphor for life in that.
3. Just look for "interesting" features, which are defined by diversity (edges or color changes, for example). Warning: This way can lead to madness—too much "interestingness" draws the network to optical illusions. So keep it sane, and emphasize the kinds of features that are likely to be real in nature, as opposed to artifacts or

errors. This is called "regularization," and there are lots of techniques for this, such as whether those kinds of features have been
seen before (learned) or are too "high frequency" (like static)
rather than "low frequency" (more continuous, like actual real-
world features).

Just because AI systems sometimes end up in local minima, don't
conclude that this makes them any less like life. Humans—indeed, probably all life-forms—are often stuck in local minima.

Take our understanding of the game of Go, which was taught and
learned and optimized by humans for thousands of years. It took AIs
less than three years to find out that we'd been playing it wrong all along
and that there were better, almost alien, solutions to the game that we'd
never considered—mostly because our brains don't have the processing
power to consider so many moves ahead.

Even in chess, which is ten times easier and was thought to be understood, brute-force machines could beat us at our own strategies. Chess,
too, turned out, when explored by superior neural-network AI systems, to
have weird but superior strategies we'd never considered, like sacrificing
queens early to gain an obscure long-term advantage. It's as if we had been
playing 2D versions of games that actually existed in higher dimensions.

If any of this sounds familiar, it's because physics has been wrestling
with these sorts of topological problems for decades. The notion of space
being many-dimensional, and math reduced to understanding the geom-
etries and interactions of "membranes" beyond the reach of our senses,
is where Grand Unified Theorists go to die. But unlike multidimensional
theoretical physics, AI is something we can actually experiment with
and measure.

So that's what we're going to do. The next few decades will be an ex-
plosive exploration of ways to think that 7 million years of evolution never
found. We're going to rock ourselves out of local minima and find deeper
minima, maybe even global minima. And when we're done, we may even
have taught machines to seem as smart as a mosquito, forever descending
the cosmic gradients to an ultimate goal, whatever that may be.

Chapter 15

"INFORMATION" FOR WIENER, FOR SHANNON, AND FOR US

DAVID KAISER

David Kaiser is Germeshausen Professor of the History of Science and professor of physics at MIT, and head of its Program in Science, Technology and Society. He is the author of How the Hippies Saved Physics: Science, Counterculture, and the Quantum Revival *and* American Physics and the Cold War Bubble *(forthcoming).*

David Kaiser is a physicist atypically interested in the intersection of his science with politics and culture, about which he has written widely.

In the first meeting (in Washington, Connecticut) that preceded the crafting of this book, he commented on the change in how "information" is viewed since Wiener's time: the military-industrial Cold War era. Back then, Wiener compared information, metaphorically, to entropy, in that it could not be conserved—i.e., monopolized; thus, he argued, our atomic secrets and other such classified matters would not remain secrets for long. Today, whereas (as Wiener might have expected) information, fake or not, is leaking all over the other Washington, information in the economic world has indeed been stockpiled, commodified, and monetized.

This lockdown, David said, was "not all good, not all bad"—depending, I guess, on whether you're sick of being pestered by ads for socks or European river cruises popping up in your browser minutes after you've bought them.

To say nothing of information's proliferation. David complained to the rest of us attending the meeting that in Wiener's time, physicists could "take the entire *Physical Review*. It would sit comfortably in front of us in a manageable pile. Now we're awash in fifty thousand open-source journals per minute," full of God-knows-what. Neither of these developments would Wiener have anticipated, said David, prompting him to ask, "Do we need a new set of guiding metaphors?"

n *The Sleepwalkers*, a sweeping history of scientific thought from ancient times through the Renaissance, Arthur Koestler identified a tension that has marked the most dramatic leaps of our cosmological imagination. In reading the great works of Nicolaus Copernicus and Johannes Kepler today, Koestler argued, we are struck as much by their strange unfamiliarity—their embeddedness in the magic or mysticism of an earlier age—as by their modern-sounding insights.

I detect that same doubleness—the zigzag origami folds of old and new—in Norbert Wiener's classic *The Human Use of Human Beings*. First published in 1950 and revised in 1954, the book is in many ways extraordinarily prescient. Wiener, the MIT polymath, recognized before most observers that "society can only be understood through a study of the messages and the communication facilities which belong to it." Wiener argued that feedback loops, the central feature of his theory of cybernetics, would play a determining role in social dynamics. Those loops would not only connect people with one another but connect people with machines, and—crucially—machines with machines.

Wiener glimpsed a world in which information could be separated from its medium. People, or machines, could communicate patterns across vast distances and use them to fashion new items at the endpoints, without "moving a . . . particle of matter from one end of the line to the other," a vision now realized in our world of networked 3D printers.

Wiener also imagined machine-to-machine feedback loops driving huge advances in automation, even for tasks that had previously relied on human judgment. "The machine plays no favorites between manual labor and white-collar labor," he observed.

For all that, many of the central arguments in *The Human Use of Human Beings* seem closer to the 19th century than the 21st. In particular, although Wiener made reference throughout to Claude Shannon's then-new work on information theory, he seems not to have fully embraced Shannon's notion of information as consisting of irreducible, meaning-free bits. Since Wiener's day, Shannon's theory has come to undergird recent advances in "Big Data" and "deep learning," which makes it all the more interesting to revisit Wiener's cybernetic imagination. How might tomorrow's artificial intelligence be different if practitioners were to reinvest in Wiener's guiding vision of "information"?

When Wiener wrote *The Human Use of Human Beings*, his experiences of war-related research, and of what struck him as the moral ambiguities of intellectual life amid the military-industrial complex, were still fresh. Just a few years earlier, he had announced in the pages of the *Atlantic Monthly* that he would not "publish any future work of mine which may do damage in the hands of irresponsible militarists."* He remained ambivalent about the transformative power of new technologies, indulging in neither the boundless hype nor the digital utopianism of later pundits.

"Progress imposes not only new possibilities for the future but new restrictions," he wrote in *Human Use*. He was concerned about human-made restrictions as well as technological ones, especially Cold War restrictions that threatened the flow of information so critical to cybernetic systems: "Under the impetus of Senator [Joseph] McCarthy and his imitators, the blind and excessive classification of military information" was driving political leaders in the United States to adopt a "secretive frame of

---

* Norbert Wiener, "A Scientist Rebels," *Atlantic Monthly*, January 1947.

mind paralleled in history only in the Venice of the Renaissance." Wiener, echoing many outspoken veterans of the Manhattan Project, argued that the postwar obsession with secrecy—especially around nuclear weapons—stemmed from a misunderstanding of the scientific process. The only genuine secret about the production of nuclear weapons, he wrote, was whether such bombs could be built. Once that secret had been revealed, with the bombings of Hiroshima and Nagasaki, no amount of state-imposed secrecy would stop others from puzzling through chains of reasoning like those the Manhattan Project researchers had followed. As Wiener memorably put it, "There is no Maginot Line of the brain."

To drive this point home, Wiener borrowed Shannon's fresh ideas about information theory. In 1948, Shannon, a mathematician and engineer working at Bell Labs, had published a pair of lengthy articles in the *Bell System Technical Journal*. Introducing the new work to a broad readership in 1949, mathematician Warren Weaver explained that in Shannon's formulation, "the word *information* . . . is used in a special sense that must not be confused with its ordinary usage. In particular, *information* must not be confused with meaning."* Linguists and poets might be concerned about the "semantic" aspects of communication, Weaver continued, but not engineers like Shannon. Rather, "this word 'information' in communication theory relates not so much to what you *do* say, as to what you *could* say." In Shannon's now-famous formulation, the information content of a string of symbols was given by the logarithm of the number of possible symbols from which a given string was chosen. Shannon's key insight was that the information of a message was just like the entropy of a gas: a measure of the system's disorder.

Wiener borrowed this insight when composing *Human Use*. If information was like entropy, then it could not be conserved—or contained. Physicists in the 19th century had demonstrated that the total energy of a physical system must always remain the same, a perfect balance between the start and the end of a process. Not so for entropy, which would

---

* Warren Weaver, "Recent Contributions to the Mathematical Theory of Communication," in Claude Shannon and Warren Weaver, *The Mathematical Theory of Communication* (Urbana: University of Illinois Press, 1949), 8 (emphasis in original). Shannon's 1948 papers were republished in the same volume.

inexorably increase over time, an imperative that came to be known as the second law of thermodynamics. From that stark distinction—energy is conserved, whereas entropy must grow—followed enormous cosmic consequences. Time must flow forward; the future cannot be the same as the past. The universe could even be careening toward a "heat death," some far-off time when the total stock of energy had uniformly dispersed, achieving a state of maximum entropy, after which no further change could occur.

If information *qua* entropy could not be conserved, then Wiener concluded it was folly for military leaders to try to stockpile the "scientific know-how of the nation in static libraries and laboratories." Indeed, "no amount of scientific research, carefully recorded in books and papers, and then put into our libraries with labels of secrecy, will be adequate to protect us for any length of time in a world where the effective level of information is perpetually advancing." Any such efforts at secrecy, classification, or the containment of information would fail, Wiener argued, just as surely as hucksters' schemes for perpetual-motion machines faltered in the face of the second law of thermodynamics.

Wiener criticized the American "orthodoxy" of free-market fundamentalism in much the same way. For most Americans, "questions of information will be evaluated according to a standard American criterion: a thing is valuable as a commodity for what it will bring in the open market." Indeed, "[T]he fate of information in the typically American world is to become something which can be bought or sold"; most people, he observed, "cannot conceive of a piece of information without an owner." Wiener considered this view to be as wrong-headed as rampant military classification. Again he invoked Shannon's insight: Since "information and entropy are not conserved," they are "equally unsuited to being commodities."

Information cannot be conserved—so far, so good. But did Wiener really have Shannon's "information" in mind? The crux of Shannon's argument, as Weaver had emphasized, was to distinguish a colloquial sense of "information," as message with meaning, from an abstracted, rarefied notion of strings of symbols arrayed with some probability and selected from an enormous universe of gibberish. For Shannon, "information" could be quantified because its fundamental unit, the bit, was a unit of conveyance rather than understanding.

When Wiener characterized "information" throughout *Human Use*, on the other hand, he tilted time and again to a classical, humanistic sense of the term. "A piece of information," he wrote—tellingly, not a "bit" of information—"in order to contribute to the general information of the community, must say something substantially different from the community's previous common stock of information." This was why "schoolboys do not like Shakespeare," he concluded: The Bard's couplets may depart starkly from random bitstreams, but they had nonetheless become all too familiar to the sense-making public and "absorbed into the superficial clichés of the time."

At least the information content of Shakespeare had once seemed fresh. During the postwar boom years, Wiener fretted, the "enormous per capita bulk of communication"—ranging across newspapers and movies to radio, television, and books—had bred mediocrity, an informational reversion to the mean. "More and more we must accept a standardized inoffensive and insignificant product which, like the white bread of the bakeries, is made rather for its keeping and selling properties than for its food value." "Heaven save us," he pleaded, "from the first novels which are written because a young man desires the prestige of being a novelist rather than because he has something to say! Heaven save us likewise from the mathematical papers which are correct and elegant but without body or spirit." Wiener's treatment of "information" sounded more like Matthew Arnold in 1869* than Claude Shannon in 1948—more "body and spirit" than "bit." Wiener shared Arnold's Romantic view of the "content producer" as well: "Properly speaking the artist, the writer, and the scientist should be moved by such an irresistible impulse to create that, even if they were not being paid for their

* Matthew Arnold, *Culture and Anarchy*, ed. Jane Garnett (Oxford, UK: Oxford University Press, 2006).

work, they would be willing to pay to get the chance to do it." *L'art pour l'art*, that 19th-century cry: Artists should suffer for their work; the quest for meaningful expression should always trump lucre.

To Wiener, this was the proper measure of "information": body, spirit, aspiration, expression. Yet to argue against its commodification, Wiener reverted again to Shannon's mathematics of information-as-entropy.

F lash forward to our day. In many ways, Wiener has been proved right. His vision of networked feedback loops driven by machine-to-machine communication has become a mundane feature of everyday life. From the earliest stirrings of the Internet Age, moreover, digital piracy has upended the view that "information"—in the form of songs, movies, books, or code—could remain contained. Put up a paywall here, and the content will diffuse over there, all so much informational entropy that cannot be conserved.

On the other hand, enormous multinational corporations—some of the largest and most profitable in the world—now routinely disprove Wiener's contention that "information" cannot be stockpiled or monetized. Ironically, the "information" they trade in is closer to Shannon's definition than Wiener's, Shannon's mathematical proofs notwithstanding.

While Google Books may help circulate hundreds of thousands of works of literature for free, Google itself—like Facebook, Amazon, Twitter, and their many imitators—has commandeered a baser form of "information" and exploited it for extraordinary profit. Petabytes of Shannon-like information—a seemingly meaningless stream of clicks, "likes," and retweets, collected from virtually every person who has ever touched a networked computer—are sifted through proprietary "deep-learning" algorithms to microtarget everything from the advertisements we see to the news stories (fake or otherwise) we encounter while browsing the Web.

Back in the early 1950s, Wiener had proposed that researchers study the structures and limitations of ants—in contrast to humans—so that machines might one day achieve the "almost indefinite intellectual expansion" that people (rather than insects) can attain. He found solace in the notion that machines could come to dominate us only "in the last stages of increasing entropy," when "the statistical differences among individuals are nil." Today's data-mining algorithms turn Wiener's approach on its head. They produce profit by exploiting our reptilian brains rather than imitating our cerebral cortexes, harvesting information from all our late-night, blog-addled, pleasure-seeking clickstreams—leveraging precisely the tiny, residual "statistical differences among individuals."

To be sure, some recent achievements in artificial intelligence have been remarkably impressive. Computers can now produce visual artworks and musical compositions akin to those of recognized masters, creating just the sort of "information" that Wiener most prized. But by far the largest impact on society to date has come from the collection and manipulation of Shannon-like information, which has reshaped our shopping habits, political participation, personal relationships, expectations of privacy, and more.

What might "deep learning" evolve into if the fundamental currency becomes "information" as Wiener defined it? How might the field shift if reanimated by Wiener's deep moral convictions, informed as they were by his prescient concerns about rampant militarism, runaway corporate profit seeking, the self-limiting features of secrecy, and the reduction of human expression to interchangeable commodities? Perhaps "deep learning" might then become the cultivation of meaningful information rather than the relentless pursuit of potent, if meaningless, bits.

Chapter 16

SCALING

NEIL GERSHENFELD

Neil Gershenfeld is a physicist and director of MIT's Center for Bits and Atoms. He is the author of FAB, *co-author (with Alan Gershenfeld and Joel Cutcher-Gershenfeld) of* Designing Reality, *and founder of the global Fab Lab network.*

In the aforementioned Connecticut discussion on *The Human Use of Human Beings*, **Neil Gershenfeld** provided some fresh air, of a kind, by professing that he hated the book. His remark was met by universal laughter—as was his observation that computer science was one of the worst things to happen to computers, or to science. His overall contention was that Wiener missed the implications of the digital revolution that was happening around him—although some would say this charge can't be leveled at someone on the ground floor and lacking clairvoyance.

"The tail wagging the dog of my life," he told us, "has been Fab Labs and the maker movement, and [when] Wiener talks about the threat of automation he misses the inverse, which is that access to the means for automation can empower people, and in Fab Labs, the corner I've been involved in, that's an exponential."

In 2003, I visited Neil at MIT, where he runs the Center for Bits and Atoms. Hours later, I emerged from what had been an exuberant display of very weird stuff. He showed me the work of one student in his popular rapid-prototyping class (How to Make Almost Anything), a sculptor with no engineering background who had made a portable personal space for screaming that saves up your screams and plays them back later. Another student in the class had made a Web browser that lets parrots navigate the net. Neil himself was doing fundamental research on the road map to that sci-fi staple, a "universal replicator." It was a visit that took me a couple of years to get my head around.

Neil manages a global network of Fab Labs—small-scale manufacturing systems, enabled by digital technologies, that give people the wherewithal to build whatever they'd like. As guru of the maker movement, which merges digital communication and computation with fabrication, he sometimes feels outside the current heated debate on AI safety. "My ability to do research rests on tools that augment my capabilities," he says. "Asking whether or not they are intelligent is as fruitful as asking how I know I exist—amusing philosophically, but not testable empirically." What interests him is "how bits and atoms relate—the boundary between digital and physical. Scientifically, it's the most exciting thing I know."

D iscussions about artificial intelligence have been oddly ahistorical. They could better be described as manic-depressive; depending on how you count, we're now in the fifth boom-bust cycle. Those swings mask the continuity in the underlying progress and the implications for where it's headed.

The cycles have come in roughly decade-long waves. First there were mainframes, which by their very existence were going to automate away work. That ran into the reality that it was hard to write programs to do tasks that were simple for people to do. Then came expert systems, which were going to codify and then replace the knowledge of experts. These ran into difficulty in assembling that knowledge and reasoning about cases not already covered. Perceptrons sought to get around these problems by modeling how the brain learns, but they were unable to do much of anything. Multilayer perceptrons could handle test problems that had tripped up those simpler networks, but their demonstrations did poorly on unstructured, real-world problems. We're now in the deep-learning era, which is delivering on many of the early AI promises but in a way that's considered hard to understand, with consequences ranging from intellectual to existential threats.

Each of these stages was heralded as a revolutionary advance over the limitations of its predecessors, yet all effectively do the same thing: They make inferences from observations. How these approaches relate can be understood by how they scale—that is, how their performance depends on the difficulty of the problem they're addressing. Both a light switch and a self-driving car must determine their operators' intentions, but the former has just two options to choose from, whereas the latter has many more. The AI-boom phases have started with promising examples in limited domains; the bust phases came with the failure of those demonstrations to handle the complexity of less-structured, practical problems.

Less apparent is the steady progress we've made in mastering scaling. This progress rests on the technological distinction between linear and exponential functions—a distinction that was becoming evident at the dawn of AI but with implications for AI that weren't appreciated until many years later.

In one of the founding documents of the study of intelligent machines, *The Human Use of Human Beings*, Norbert Wiener does a remarkable job of identifying many of the most significant trends to arise since he wrote it, along with noting the people responsible for them and then consistently failing to recognize why these people's work proved to be so important. Wiener is credited with creating the field of cybernetics; I've never understood what that is, but what's missing from the book is at the heart of how AI has progressed. This history matters because of the echoes of it that persist to this day.

Claude Shannon makes a cameo appearance in the book, in the context of his thoughts about the prospects for a chess-playing computer. Shannon was doing something much more significant than speculating at the time: He was laying the foundations for the digital revolution. As a graduate student at MIT, he worked for Vannevar Bush on the Differential Analyzer. This was one of the last great analog computers, a room full of gears and shafts. Shannon's frustration with the difficulty of solving problems this way led him in 1937 to write what might be the best master's thesis ever. In it, he showed how electrical circuits could be designed to evaluate arbitrary logical expressions, introducing the basis for universal digital logic.

After MIT, Shannon studied communications at Bell Labs. Analog telephone calls degraded with distance; the farther they traveled, the worse they sounded. Rather than continue to improve them incrementally, Shannon showed in 1948 that by communicating with symbols rather than continuous quantities, the behavior is very different. Converting speech waveforms to the binary values of 1 and 0 is an example, but many other sets of symbols can be (and are) used in digital communications. What matters is not the particular symbols but rather the ability to detect and correct errors. Shannon found that if the noise is above a threshold (which depends on the system design), then there are certain to be errors. But if the noise is below a threshold, then a linear increase in the physical resources representing the symbol results in an exponential decrease in the likelihood of making an error in correctly receiving the symbol. This relationship was the first of what we'd now call a threshold theorem.

Such scaling falls off so quickly that the probability of an error can be so small as to effectively never happen. Each symbol sent multiplies rather than adds to the certainty, so that the probability of a mistake can go from 0.1 to 0.01 to 0.001, and so forth. This exponential decrease in communication errors made possible an exponential increase in the capacity of communication networks. And that eventually solved the problem of where the knowledge in an AI system came from.

For many years, the fastest way to speed up a computation was to do nothing—just wait for computers to get faster. In the same way, there were years of AI projects that aimed to accumulate everyday knowledge by laboriously entering pieces of information. That didn't scale; it could progress only as fast as the number of people doing the entering. But when phone calls, newspaper stories, and mail messages all moved onto the Internet, everyone doing any of those things became a data generator. The result was an exponential rather than a linear rate of knowledge accumulation.

John von Neumann also has a cameo in *The Human Use of Human Beings*, for game theory. What Wiener missed here was von Neumann's seminal role in digitizing computation. Whereas analog communication degraded with distance, analog computing (like the Differential Analyzer) degraded with time, accumulating errors as it progressed. Von Neumann presented in 1952 a result corresponding to Shannon's for computation (they had met at the Institute for Advanced Study in Princeton), showing that it was possible to compute reliably with an unreliable computing device by using symbols rather than continuous quantities. This was, again, a scaling argument, with a linear increase in the physical resources representing the symbol resulting in an exponential reduction in the error rate as long as the noise was below a threshold. That's what makes it possible to have a billion transistors in a computer chip, with the last one as useful as the first one. This relationship led to an exponential increase in computing performance, which solved a second problem in AI: how to process exponentially increasing amounts of data.

The third problem that scaling solved for AI was coming up with the rules for reasoning without having to hire a programmer for each problem. Wiener recognized the role of feedback in machine learning, but he missed the key role of representation. It's not possible to store all possible images in a self-driving car, or all possible sounds in a conversational computer; they have to be able to generalize from experience. The "deep" part of deep learning refers not to the (hoped-for) depth of insight but to the depth of the mathematical network layers used to make predictions. It turned out that a linear increase in network complexity led to an exponential increase in the expressive power of the network.

If you lose your keys in a room, you can search for them. If you're not sure which room they're in, you have to search all the rooms in a building. If you're not sure which building they're in, you have to search all the rooms in all the buildings in a city. If you're not sure which city they're in, you have to search all the rooms in all the buildings in all the cities. In AI, finding the keys corresponds to things like a car safely following the road, or a computer correctly interpreting a spoken command, and the rooms and buildings and cities correspond to all of the options that have to be considered. This is called the curse of dimensionality.

The solution to the curse of dimensionality came in using information about the problem to constrain the search. The search algorithms themselves are not new. But when applied to a deep-learning network, they adaptively build up representations of where to search. The price of this is that it's no longer possible to exactly solve for the best answer to a problem, but typically all that's needed is an answer that's good enough.

Taken together, it shouldn't be surprising that these scaling laws have allowed machines to become effectively as capable as the corresponding stages of biological complexity. Neural networks started out with a goal of modeling how the brain works. That goal was abandoned as they evolved into mathematical abstractions unrelated to how neurons actually function. But now there's a kind of convergence that can be thought of as forward- rather than reverse-engineering biology, as the results of deep learning echo brain layers and regions.

One of the most difficult research projects I've managed paired what we'd now call data scientists with AI pioneers. It was a miserable experience in moving goalposts. As the former progressed in solving long-standing problems posed by the latter, this was deemed to not count because it wasn't accompanied by corresponding leaps in understanding the solutions. What's the value of a chess-playing computer if you can't explain how it plays chess?

The answer, of course, is that it can play chess. There is interesting emerging research that is applying AI to AI—that is, training networks to explain how they operate. But both brains and computer chips are hard to understand by watching their inner workings; they're easily interpreted only by observing their external interfaces. We come to trust (or not) brains and computer chips alike based on experience that tests them rather than on explanations for how they work.

Many branches of engineering are making a transition from what's called imperative to declarative or generative design. This means that instead of explicitly designing a system with tools like CAD files, circuit schematics, and computer code, you describe what you want the system to do and then an automated search is done for designs that satisfy your goals and restrictions. This approach becomes necessary as design complexity exceeds what can be understood by a human designer. While that might sound like a risk, human understanding comes with its own limits; engineering design is littered with what appeared to be good insights that have had bad consequences. Declarative design rests on all the advances in AI, plus the improving fidelity of simulations to virtually test designs.

The mother of all design problems is the one that resulted in us. The way we're designed resides in one of the oldest and most conserved parts of the genome, called the Hox genes. These are genes that regulate genes, in what are called developmental programs. Nothing in your genome stores the design of your body; your genome stores, rather, a series of steps to follow that results in your body. This is an exact parallel to how search is done in AI. There are too many possible body plans to search over, and most modifications would be either inconsequential or fatal. The Hox genes are a representation of a productive place for evolutionary search. It's a kind of natural intelligence at the molecular level.

AI has a mind-body problem in that it has no body. Most work on AI is done in the cloud, running on virtual machines in computer centers where data are funneled. Our own intelligence is the result of a search algorithm (evolution) that was able to change our physical form as well as our programming—those are inextricably linked. If the history of AI can be understood as the working of scaling laws rather than a succession of fashions, then its future can be seen in the same way. What's now being digitized, after communication and computation, is fabrication, bringing the programmability of bits to the world of atoms. By digitizing not just designs but the construction of materials, the same lessons that von Neumann and Shannon taught us apply to exponentially increasing fabricational complexity.

I've defined digital materials to be those constructed from a discrete set of parts reversibly joined with a discrete set of relative positions and orientations. These attributes allow the global geometry to be determined from local constraints, assembly errors to be detected and corrected, heterogeneous materials to be joined, and structures to be disassembled rather than disposed of when they're no longer needed. The amino acids that are the foundation of life and the Lego bricks that are the foundation of play share these properties.

What's interesting about amino acids is that they're not interesting. They have attributes that are typical but not unusual, such as attracting or repelling water. But just twenty types of them are enough to make you. In the same way, twenty or so types of digital-material part types—conducting, insulating, rigid, flexible, magnetic, etc.—are enough to assemble the range of functions that go into making modern technologies like robots and computers.

The connection between computation and fabrication was foreshadowed by the very pioneers whose work the edifice of computing is based on. Wiener hinted at this by linking material transportation with message transportation. John von Neumann is credited with modern computer architecture, something he actually wrote very little about; the final thing he studied, and wrote about beautifully and at length, was self-reproducing systems. As an abstraction of life, he modeled a machine that can communicate a computation that constructs itself. And the final thing Alan Turing, who is credited with the theoretical framework for computer science, studied was how the instructions in genes can give rise to physical forms. These questions address a topic absent from a typical computer-science education: the physical configuration of a computation.

Von Neumann and Turing posed their questions as theoretical studies, because it was beyond the technology of their day to realize them. But with the convergence of communication and computation with fabrication, these investigations are now becoming accessible experimentally. Making an assembler that can assemble itself from the parts that it's assembling is a focus of my lab, along with collaborations to develop synthetic cells.

The prospect of physically self-reproducing automata is potentially much scarier than fears of out-of-control AI, because it moves the intelligence out here to where we live. It could be a road map leading to *Terminator*'s Skynet robotic overlords. But it's also a more hopeful prospect, because an ability to program atoms as well as bits enables designs to be shared globally while locally producing things like energy, food, and shelter—all of these are emerging as exciting early applications of digital fabrication. Wiener worried about the future of work, but he didn't question implicit assumptions about the nature of work that are challenged when consumption can be replaced by creation.

History suggests that neither utopian nor dystopian scenarios prevail; we generally end up muddling along somewhere in between. But history also suggests that we don't have to wait on history. Gordon Moore in 1965 was able to use five years of the doubling of the specifications of integrated circuits to project what turned out to be fifty years of exponential improvements in digital technologies. We've spent many of those years responding to, rather than anticipating, its implications. We have more data available now than Gordon Moore did to project fifty years of doubling the performance of digital fabrication. With the benefit of hindsight, it should be possible to avoid the excesses of digital computing and communications this time around, and, from the outset, address issues like access and literacy.

If the maker movement is the harbinger of a third digital revolution, the success of AI in meeting many of its own early goals can be seen as the crowning achievement of the first two digital revolutions. Although machine making and machine thinking might appear to be unrelated trends, they lie in each other's futures. The same scaling trends that have made AI possible suggest that the current mania is a phase that will pass, to be followed by something even more significant: the merging of artificial and natural intelligence.

It was an advance for atoms to form molecules, molecules to form organelles, organelles to form cells, cells to form organs, organs to form organisms, organisms to form families, families to form societies, and societies to form civilizations. This grand evolutionary loop can now be closed, with atoms arranging bits arranging atoms.

Chapter 17

THE FIRST MACHINE INTELLIGENCES

W. DANIEL HILLIS

W. Daniel "Danny" Hillis *is an inventor, entrepreneur, and computer scientist, Judge Widney Professor of Engineering and Medicine at USC, and author of* The Pattern on the Stone: The Simple Ideas That Make Computers Work.

While **Danny Hillis** was an undergraduate at MIT, he built a computer out of Tinkertoys. It has around ten thousand wooden parts, plays tic-tac-toe, and never loses; it's now in the Computer History Museum in Mountain View, California.

As a graduate student at the MIT Computer Science and Artificial Intelligence Laboratory in the early 1980s, Danny designed a massively parallel computer with sixty-four thousand processors. He named it the Connection Machine and founded what might have been the first AI company—Thinking Machines Corporation—to produce and market it. This was despite a lunch he had with Richard Feynman at which the celebrated physicist remarked, "That is positively the dopiest idea I ever heard." Maybe "despite" is the wrong word, since Feynman had a well-known predilection for playing with dopey ideas. In the event, he showed up on the day the company was incorporated and stayed on, for summer jobs and special assignments, to make invaluable contributions to its work.

Danny has since established a number of technology companies, of which the latest is Applied Invention, which partners with commercial enterprises to develop technological solutions to their most intractable problems. He holds hundreds of U.S. patents, covering parallel computers, touch interfaces, disk arrays, forgery prevention methods, and a slew of electronic and mechanical devices. His imagination is apparently boundless, and here he sketches some possible scenarios that will result from our pursuit of a better and better AI.

"Our thinking machines are more than metaphors," he says. "The question is not 'Will they be powerful enough to hurt us?' (they will), or whether they will always act in our best interests (they won't), but whether over the long term they can help us find our way—where we come out on the Panacea/Apocalypse continuum."

I have spoken of machines, but not only of machines having brains of brass and thews of iron. When human atoms are knit into an organization in which they are used, not in their full right as responsible human beings, but as cogs and levers and rods, it matters little that their raw material is flesh and blood. What is used as an element in a machine, is in fact an element in the machine. Whether we entrust our decisions to machines of metal, or to those machines of flesh and blood which are bureaus and vast laboratories and armies and corporations, we shall never receive the right answers to our questions unless we ask the right questions. . . . The hour is very late, and the choice of good and evil knocks at our door.

—NORBERT WIENER, *The Human Use of Human Beings*

Norbert Wiener was ahead of his time in recognizing the potential danger of emergent intelligent machines. I believe he was even further ahead in recognizing that the first artificial intelligences had already begun to emerge. He was correct in identifying the corporations and bureaus that he called "machines of flesh and blood" as the first intelligent machines. He anticipated the dangers of creating artificial superintelligences with goals not necessarily aligned with our own.

What is now clear, whether or not it was apparent to Wiener, is that these organizational superintelligences are not just made of humans, they are hybrids of humans and the information technologies that allow them to coordinate. Even in Wiener's time, the "bureaus and vast laboratories and armies and corporations" could not operate without telephones, telegraphs, radios, and tabulating machines. Today they could not operate without networks of computers, databases, and decision support systems. These hybrid intelligences are technologically augmented networks of humans. These artificial intelligences have superhuman powers. They can know more than individual humans; they can sense more; they can make more complicated analyses and more complex plans. They can have vastly more resources and power than any single individual.

Although we do not always perceive it, hybrid superintelligences such as nation-states and corporations have their own emergent goals. Although they are built by and for humans, they often act like independent intelligent entities, and their actions are not always aligned with the interests of the people who created them. The state is not always for the citizen, nor the company for the shareholder. Nor do not-for-profits, religious orders, or political parties always act in furtherance of their founding principles. Intuitively, we recognize that their actions are guided by internal goals, which is why we personify them, both legally and in our habits of thought. When talking about "what China wants" or "what General Motors is trying to do," we are not speaking in metaphors. These organizations act as intelligences that perceive, decide, and act. Like the goals of individual humans, the goals of organizations are complex and often self-contradictory, but they are true goals in the sense that they direct action. Those goals depend somewhat on the goals of the people within the organization, but they are not identical.

Any American knows how loose the tie is between the actions of the U.S. government and the diverse and often contradictory aims of its citizens. That is also true of corporations. For-profit corporations nominally serve multiple constituencies, including shareholders, senior executives, employees, and customers. These corporations differ in how they balance their loyalties and often behave in ways that serve none of their constituents. The "neurons" that carry their corporate thought are not just the human employees or the technologies that connect them; they are also coded into the policies, incentive structures, culture, and procedural habits of the corporation. The emergent corporate goals do not always reflect the values of the people who implement them. For instance, an oil company led and staffed by people who care about the environment may have incentive structures or policies that cause it to compromise environmental safety for the sake of corporate earnings. The components' good intentions are not a guarantee of the emergent system's good behavior.

Governments and corporations, both built partly of humans, are naturally motivated to at least appear to share the goals of the humans they depend upon. They could not function without the people, so they need to keep them cooperative. When such organizations appear to behave altruistically, this is often part of their motive. I once complimented the CEO of a large corporation on the contribution his company made toward a humanitarian relief effort. The CEO responded, without a trace of irony, "Yes. We have decided to do more things like that to make our brand more likable." Individuals who compose a hybrid superintelligence may occasionally exert a "humanizing" influence—for example, an employee may break company policies to accommodate the needs of another human. The employee may act out of true human empathy, but we should not attribute any such empathy to the superintelligence itself. These hybrid machines have goals, and their citizens/customers/employees are some of the resources they use to accomplish them.

We are close to being able to build superintelligences out of pure information technology, without human components. This is what people normally refer to as "artificial intelligence," or AI. It is reasonable to ask what the attitudes of the hypothetical machine superintelligences will be toward humans. Will they, too, see humans as useful resources and a good relationship with us as worth preserving? Will they be constructed to have goals that are aligned with our own? Will a superintelligence even see these questions as important? What are the "right questions" that we should be asking? I believe that one of the most important is this: What relationship will various superintelligences have to one another?

It is interesting to consider how the hybrid superintelligences currently deal with conflicts among themselves. Today, much of the ultimate power rests in the nation-states, which claim authority over a patch of ground. Whether they are optimized to act in the interests of their citizens or those of a despotic ruler, nation-states assert priority over other intelligences' desires or goals within their geographic dominion. They claim a monopoly on the use of force and recognize only other nation-states as peers. They are willing, if necessary, to demand great sacrifices of their citizens to enforce their authority, even to the point of sacrificing their citizens' lives.

This geographical division of authority made logical sense when most of the actors were humans who spent their lives within a single nation-state, but now that the actors of importance include geographically distributed hybrid intelligences such as multinational corporations, that logic is less obvious. Today we live in a complex transitional period, when distributed superintelligences still largely rely on the nation-states to settle the arguments arising among them. Often, those arguments are resolved differently in different jurisdictions. It is becoming more difficult even to assign individual humans to nation-states: International travelers living and working outside their native countries, refugees, and immigrants (documented and not) are still dealt with as awkward exceptions. Superintelligences built purely of information technology will prove even more awkward for the territorial system of authority, since there is no reason why they need to be tied to physical resources in a single country—or even to any particular physical resources at all. An artificial intelligence might well exist "in the cloud" rather than at any physical location.

I can imagine at least four scenarios for how machine superintelligences will relate to hybrid superintelligences.

In one obvious scenario, multiple machine intelligences will ultimately be controlled by, and allied with, individual nation-states. In this state/AI scenario, one can envision American and Chinese super-AIs wrestling each other for resources on behalf of their state. In some sense, these AIs would be citizens of their nation-state in the way that many commercial corporations often act as "corporate citizens" today. In this scenario, the host nation-states would presumably give the machine superintelligences the resources they need to work for the state's advantage. Or, to the degree that the superintelligences can influence their state governments, they will presumably do so to enhance their own power, for instance by garnering a larger share of the state's resources. Nation-states' AIs might not want competing AIs to grow up within their jurisdiction. In this scenario, the superintelligences become an extension of the state, and vice versa.

The state/AI scenario seems plausible, but it is not our current course. Our most powerful and rapidly improving artificial intelligences are controlled by for-profit corporations. This is the corporate/AI scenario, in which the balance of power between nation-states and corporations becomes inverted. Today, the most powerful and intelligent collections of machines are probably owned by Google, but companies like Amazon, Baidu, Microsoft, Facebook, Apple, and IBM may not be far behind. These companies all see a business imperative to build artificial intelligences of their own. It is easy to imagine a future in which corporations independently build their own machine intelligences, protected within firewalls preventing the machines from taking advantage of one another's knowledge. These machines will be designed to have goals aligned with those of the corporation. If this alignment is effective, nation-states may continue to lag behind in developing their own artificial-intelligence capability and instead depend on their "corporate citizens" to do it for them. To the extent that corporations successfully control the goals, they will become more powerful and autonomous than nation-states.

Another scenario, perhaps the one people fear the most, is that artificial intelligences will not be aligned with either humans or hybrid superintelligences but will act solely in their own interests. They might even merge into a single machine superintelligence, since there may be no technical requirement for machine intelligences to maintain distinct identities. The attitude of a self-interested super-AI toward hybrid superintelligences is likely to be competitive. Humans might be seen as minor annoyances, like ants at a picnic, but hybrid superintelligences—like corporations, organized religions, and nation-states—could be existential threats. Like hybrid superintelligences, AIs might see humans mostly as useful tools to accomplish their goals, as pawns in their competition with the other superintelligences. Or we might simply be irrelevant. It is not impossible that a machine intelligence has already emerged and we simply do not recognize it as such. It may not wish to be noticed, or it may be so alien to us that we are incapable of perceiving it. This makes the self-interested AI scenario the most difficult to imagine. I believe the easy-to-imagine versions, like the humanoid intelligent robots of science fiction, are the least likely. Our most complex machines, like the Internet, have already grown beyond the detailed understanding of a single human, and their emergent behaviors may be well beyond our ken.

The final scenario is that machine intelligences will not be allied with one another but instead will work to further the goals of humanity as a whole. In this optimistic scenario, AI could help us restore the balance of power between the individual and the corporation, between the citizen and the state. It could help us solve the problems that have been created by hybrid superintelligences that subvert the goals of humans. In this scenario, AIs will empower us by giving us access to processing capacity and knowledge currently available only to corporations and states. In effect, they could become extensions of our own individual intelligences, in furtherance of our human goals. They could make our weak individual intelligences strong. This prospect is both exciting and plausible. It is plausible because we have some choice in what we build, and we have a history of using technology to expand and augment our human capacities. As airplanes have given us wings and engines have given us muscles to move mountains, so our network of computers may amplify and extend our minds. We may not fully understand or control our destiny, but we have a chance to bend it in the direction of our values. The future is not something that will happen to us; it is something that we will build.

WHY WIENER SAW WHAT OTHERS MISSED

There is in electrical engineering a split which is known in Germany as the split between the technique of strong currents and the technique of weak currents, and which we know as the distinction between power and communication engineering. It is this split which separates the age just past from that in which we are now living.

—NORBERT WIENER, *Cybernetics: or Control and Communication in the Animal and the Machine*

Cybernetics is the study of the how the weak can control the strong. Consider the defining metaphor of the field: the helmsman guiding a ship with a tiller. The helmsman's goal is to control the heading of the ship, to keep it on the right course. The information, the message that is sent to the helmsman, comes from the compass or the stars, and the helmsman closes the feedback loop by sending the steering messages through the gentle force of his hand on the tiller. In this picture, we see the ship tossing in powerful wind and waves in the real world, controlled by the communication system of messages in the world of information.

Yet the distinction between "real" and "information" is mostly a difference in perspective. The signals that carry messages, like the light of the stars and the pressure of the hand on the tiller, exist in a world of energy and forces, as does the helmsman. The weak forces that control the rudder are as real and physical as the strong forces that toss the ship. If we shift our cybernetics perspective from the ship to the helmsman, the pressures on the rudder become a strong force of muscles controlled by the weak signals in the mind of the helmsman. These messages in the helmsman's mind are amplified into a physical force strong enough to steer the ship. Or instead, we can zoom out and take a large cybernetics perspective. We might see the ship itself as part of a vast trade network, part of a feedback loop that regulates the price of commodities through the flow of goods. In this perspective, the tiny ship is merely a messenger. So the distinction between the physical world and the information world is a way to describe the relationship between the weak and the strong.

Wiener chose to view the world from the vantage point and scale of the individual human. As a cyberneticist, he took the perspective of the weak protagonist embedded within a strong system, trying to make the best of limited powers. He incorporated this perspective in his very definition of information. "Information," he said, "is a name for the content of what is exchanged with the outer world as we adjust to it, and make our adjustment felt upon it." In his words, information is what we use to "live effectively within that environment."* For Wiener, information is a way for the weak to effectively cope with the strong. This viewpoint is also reflected in Gregory Bateson's definition of information as "a difference that makes a difference," by which he meant the small difference that makes a big difference.

The goal of cybernetics was to create a tiny model of the system using "weak currents" to amplify and control "strong currents" of the real world. The central insight was that a control problem could be solved by building an analogous system in the information space of messages and then amplifying solutions into the larger world of reality. Inherent in the motion of a control system is the concept of amplification, which makes the small big and the weak strong. Amplification allows the difference that makes a difference to make a difference.

In this way of looking at the world, a control system needed to be as complex as the system it controlled. Cyberneticist W. Ross Ashby proved that this was true in a precise mathematical sense, in what is now called Ashby's Law of Requisite Variety, or sometimes the First Law of Cybernetics. The law tells us that to control a system completely, the controller must be as complex as the controlled. Thus cyberneticists tended to see control systems as a kind of analog of the systems they governed, like the homunculus—the hypothetical little person inside the brain who controls the actual person.

This notion of analogous structure is sometimes confused with the notion of analog encoding of messages, but the two are logically distinct. Norbert Wiener was much impressed with Vannevar Bush's digital

---

* *The Human Use of Human Beings* (Boston: Houghton Mifflin, 1954), 17–18.

Differential Analyzer, which could be reconfigured to match the structure of whatever problem it was given to solve but used digital signal encoding. Signals could be simplified to openly represent the relevant distinctions, allowing them to be more accurately communicated and stored. In digital signals, one needed only to preserve the difference in signals that made a difference. It is this distinction and signal coding that we commonly use to distinguish "analog" versus "digital." Digital signal encoding was entirely compatible with cybernetic thinking—in fact, enabling to it. What was constraining to cybernetics was the presumption of an analogy of structure between the controller and the controlled. By the 1930s, Kurt Gödel, Alonzo Church, and Alan Turing had all described universal systems of computation, in which the computation required no structural analogy to functions that were computed. These universal computers could also compute the functions of control.

The analogy of structure between the controller and the controlled was central to the cybernetic perspective. Just as digital coding collapses the space of possible messages into a simplified version that represents only the difference that makes a difference, so the control system collapses the state space of a controlled system into a simplified model that reflects only the goals of the controller. Ashby's Law does not imply that every controller must model every state of the system but only those states that matter for advancing the controller's goals. Thus, in cybernetics, the goal of the controller becomes the perspective from which the world is viewed.

Norbert Wiener adopted the perspective of the individual human relating to vast organizations and trying to "live effectively within that environment." He took the perspective of the weak trying to influence the strong. Perhaps this is why he was able to notice the emergent goals of the "machines of flesh and blood" and anticipate some of the human challenges posed by these new intelligences—hybrid machine intelligences with goals of their own.

Chapter 18

WILL COMPUTERS BECOME
OUR OVERLORDS?

VENKI RAMAKRISHNAN

Venki Ramakrishnan is a scientist at the Medical Research Council Laboratory of Molecular Biology, Cambridge University; recipient of the Nobel Prize in Chemistry (2009); current president of the Royal Society; and the author of Gene Machine: The Race to Discover the Secrets of the Ribosome.

Venki Ramakrishnan is a Nobel Prize–winning biologist whose many scientific contributions include his work on the atomic structure of the ribosome—in effect, a huge molecular machine that reads our genes and makes proteins. His work would have been impossible without powerful computers. The Internet made his own work a lot easier and, he notes, acted as a leveler internationally: "When I grew up in India, if you wanted to get a book, it would show up six months or a year after it had already come out in the West. . . . Journals would arrive by surface mail a few months later. I didn't have to deal with it, because I left India when I was nineteen, but I know Indian scientists had to deal with it. Today they have access to information at the click of a button. More important, they have access to lectures. They can listen to Richard Feynman. That would have been a dream of mine when I was growing up. They can just watch Richard Feynman on the Web. That's a big leveling in the field." And yet . . . "Along with the benefits [of the Web], there is now a huge amount of noise. You have all of these people spouting pseudoscientific jargon and pushing their own ideas as if they were science."

As president of the Royal Society, Venki worries, too, about the broader issue of trust: public trust in evidence-based scientific findings, but also trust among scientists, bolstered by rigorous checking of one another's conclusions—trust that is in danger of eroding because of the "black box" character of deep-learning computers. "This [erosion] is going to happen more and more, as data sets get bigger, as we have genome-wide studies, population studies, and all sorts of things," he says. "How do we, as a science community, grapple with this and communicate to the public a sense of what science is about, what is reliable in science, what is uncertain in science, and what is just plain wrong in science?"

former colleague of mine, Gérard Bricogne, used to joke that carbon-based intelligence was simply a catalyst for the evolution of silicon-based intelligence. For quite a long time, both Hollywood movies and scientific Jeremiahs have been predicting our eventual capitulation to our computer overlords. We all await the singularity, which always seems to be just over the horizon.

In a sense, computers have already taken over, facilitating virtually every aspect of our lives—from banking, travel, and utilities to the most intimate personal communication. I can see and talk to my grandson in New York *for free*. I remember when I first saw the 1968 movie *2001: A Space Odyssey*, the audience laughed at the absurdly cheap cost of a picturephone call from space: $1.70, at a time when a long-distance call within the U.S. was $3 per minute.

However, the convenience and power of computers is also something of a Faustian bargain, for it comes with a loss of control. Computers prevent us from doing things we want. Try getting on a flight if you arrive at the airport and the airline computer systems are down, as happened not so long ago to British Airways at Heathrow. The planes, pilots, and passengers were all there; even the air-traffic controls were working. But no flights for that airline were allowed to take off. Computers also make us do things we *don't* want—by generating mailing lists and print labels to send us all millions of pieces of unwanted mail, which we humans have to sort, deliver, and dispose of.

But you ain't seen nothing yet. In the past, we programmed computers using algorithms we understood at least in principle. So when machines did amazing things like beating world chess champion Garry Kasparov, we could say that the victorious programs were designed with algorithms based on our own understanding—using, in this instance, the experience and advice of top grand masters. Machines were simply faster at doing brute-force calculations, had prodigious amounts of memory, and were not prone to errors. One article described Deep Blue's victory not as that of a computer, which was just a dumb machine, but as the victory of hundreds of programmers over Kasparov, a single individual.

That way of programming is changing dramatically. After a long hiatus, the power of machine learning has taken off. Much of the change came when programmers, rather than trying to anticipate and code for every possible contingency, allowed computers to train themselves on data, using deep neural networks based on models of how our own brains learn. They use probabilistic methods to "learn" from large quantities of data; computers can recognize patterns and come up with conclusions on their own. A particularly powerful method is called reinforcement learning, by which the computer learns, without prior input, which variables are important and how much to weight them to reach a certain goal. This method in some sense mimics how we learn as children. The results from these new approaches are amazing.

Such a deep-learning program was used to teach a computer to play Go, a game that only a few years ago was thought to be beyond the reach of AI because it was so hard to calculate how well you were doing. It seemed that top Go players relied a great deal on intuition and a feel for position, so proficiency was thought to require a particularly human kind of intelligence. But the AlphaGo program produced by Deep-Mind, after being trained on thousands of high-level Go games played by humans and then millions of games with itself, was able to beat the top human players in short order. Even more amazingly, the related

AlphaGo Zero program, which learned from scratch by playing itself, was stronger than the version trained initially on human games! It was as though the humans had been preventing the computer from reaching its true potential. The same method has recently been generalized: Starting from scratch, within just twenty-four hours, an equivalent Alpha-Zero chess program was able to beat today's top "conventional" chess programs, which in turn have beaten the best humans.

Progress has not been restricted to games. Computers are significantly better at image and voice recognition and speech synthesis than they used to be. They can detect tumors in radiographs earlier than most humans. Medical diagnostics and personalized medicine will improve substantially. Transportation by self-driving cars will keep us all safer, on average. My grandson may never have to acquire a driver's license, because driving a car will be like riding a horse today—a hobby for the few. Dangerous activities, such as mining, and tedious repetitive work will be done by computers. Governments will offer better targeted, more personalized and efficient public services. AI could revolutionize education by analyzing an individual pupil's needs and enabling customized teaching, so that each student could advance at an optimal rate.

Along with these huge benefits, of course, will come alarming risks. With the vast amounts of personal data, computers will learn more about us than we may know about ourselves; the question of who owns data about us will be paramount. Moreover, data-based decisions will undoubtedly reflect social biases: Even an allegedly neutral intelligent system designed to predict loan risks, say, may conclude that mere membership in a particular minority group makes you more likely to default on a loan. While this is an obvious example that we could correct, the real danger is that we are not always aware of biases in the data and may simply perpetuate them.

Machine learning may also perpetuate our own biases. When Netflix or Amazon tries to tell you what you might want to watch or buy, this is an application of machine learning. Currently such suggestions are sometimes laughable, but with time and more data they will get increasingly accurate, reinforcing our prejudices and likes and dislikes. Will we miss out on the random encounter that might persuade us to change our views by exposing us to new and conflicting ideas? Social media, given its influence on elections, is a particularly striking illustration of how the divide between people on different sides of the political spectrum can be accentuated.

We might have already reached the stage where most governments are powerless to resist the combined clout of a few powerful multinational companies that control us and our digital future. The fight between dominant companies today is really a fight for control over our data. They will use their enormous influence to prevent regulation of data, because their interests lie in unfettered control of it. Moreover, they have the financial resources to hire the most talented workers in the field, enhancing their power even further. We have been giving away valuable data for the sake of freebies like Gmail and Facebook, but as the journalist and author John Lanchester has pointed out in the *London Review of Books*, if it is free, then you are the product. Their real customers are the ones who pay them for access to knowledge about us, so that they can persuade us to buy their products or otherwise influence us. One way around the monopolistic control of data is to split the ownership of data away from firms that use them. Individuals would instead own and control access to their personal data (a model that would encourage competition, since people would be free to move their data to a company that offered better services). Finally, abuse of data is not limited to corporations: In totalitarian states, or even nominally democratic ones, governments know things about their citizens that Orwell could not have imagined. The use they make of this information may not always be transparent or possible to counter.

The prospect of AI for military purposes is frightening. One can imagine intelligent systems being designed to act autonomously based on real-time data and able to act faster than the enemy, starting catastrophic wars. Such wars may not necessarily be conventional or even nuclear wars. Given how essential computer networks are to modern society, it is much more likely that AI wars will be fought in cyberspace. The consequences could be just as dire.

Despite this loss of control, we continue to march inexorably into a world in which AI will be everywhere: Individuals won't be able to resist its convenience and power, and corporations and governments won't be able to resist its competitive advantages. But important questions arise about the future of work. Computers have been responsible for considerable losses in blue-collar jobs in the last few decades, but until recently many white-collar jobs—jobs that "only humans can do"—were thought to be safe. Suddenly that no longer appears to be true. Accountants, many legal and medical professionals, financial analysts and stockbrokers, travel agents—in fact, a large fraction of white-collar jobs—will disappear as a result of sophisticated machine-learning programs. We face a future in which factories churn out goods with very few employees and the movement of goods is largely automated, as are many services. What's left for humans to do?

In 1930—long before the advent of computers, let alone AI—John Maynard Keynes wrote, in an essay called "Economic Possibilities for Our Grandchildren," that as a result of improvements in productivity, society could produce all its needs with a fifteen-hour workweek. He also predicted, along with the growth of creative leisure, the end of money and wealth as a goal:

> We shall be able to afford to dare to assess the money-motive at its true value. The love of money as a possession—as distinguished from the love of money as a means to the enjoyments and realities of life—will be recognised for what it is, a somewhat disgusting morbidity, one of those semi-criminal, semi-pathological propensities which one hands over with a shudder to the specialists in mental disease.

Sadly, Keynes's predictions did not come true. Although productivity did indeed increase, the system—possibly inherent in a market economy—did not result in humans working much shorter hours. Rather, what happened is what the anthropologist and anarchist David

Graeber describes as the growth of "bullshit jobs."* While jobs that produce essentials like food, shelter, and goods have been largely automated away, we have seen an enormous expansion of sectors like corporate law, academic and health administration (as opposed to actual teaching, research, and the practice of medicine), "human resources," and public relations, not to mention new industries like financial services and telemarketing and ancillary industries in the so-called gig economy that serve those who are too busy doing all that additional work.

How will societies cope with technology's increasingly rapid destruction of entire professions and throwing large numbers of people out of work? Some argue that this concern is based on a false premise, because new jobs spring up that didn't exist before, but as Graeber points out, these new jobs won't necessarily be rewarding or fulfilling. During the first industrial revolution, it took almost a century before most people were better off. That revolution was possible only because the government of the time ruthlessly favored property rights over labor, and most people (and all women) did not have the vote. In today's democratic societies, it is not clear that the population will tolerate such a dramatic upheaval of society based on the promise that "eventually" things will get better.

Even that rosy vision will depend on a radical shake-up of education and lifelong learning. The Industrial Revolution did trigger enormous social change of this kind, including a shift to universal education. But it will not happen unless we make it happen: This is essentially about power, agency, and control. What's next for, say, the forty-year-old taxi driver or truck driver in an era of autonomous vehicles?

One idea that has been touted is that of a universal basic income, which will allow citizens to pursue their interests, retrain for new occupations, and generally be free to live a decent life. However, market economies, which are predicated on growing consumer demand over all else, may not tolerate this innovation. There is also a feeling among many that meaningful work is essential to human dignity and fulfillment. So

---

* https://strikemag.org/bullshit-jobs.

another possibility is that the enormous wealth generated by increased productivity due to automation could be redistributed to jobs requiring human labor and creativity in fields such as the arts, music, social work, and other worthwhile pursuits. Ultimately, which jobs are rewarding or productive and which are "bullshit" is a matter of judgment and may vary from society to society, as well as over time.

So far, I've focused on AI's practical consequences. As a scientist, what bothers me is our potential loss of understanding. We are now accumulating data at an incredible rate. In my own lab, an experiment generates more than a terabyte of data a day. These data are massaged, analyzed, and reduced until there is an interpretable result. But in all of this data analysis, we believe we know what's happening. We know what the programs are doing because we designed the algorithms at their heart. So when our computers generate a result, we feel that we intellectually grasp it.

The new machine-learning programs are different. Having recognized patterns via deep neural networks, they come up with conclusions, and we have no idea exactly how. When they uncover relationships, we don't understand it in the same way as if we had deduced those relationships ourselves using an underlying theoretical framework. As data sets become larger, we won't be able to analyze them ourselves even with the help of computers; rather, we will rely entirely on computers to do the analysis for us. So if someone asks us how we know something, we will simply say it is because the machine analyzed the data and produced the conclusion.

One day a computer may well come up with an entirely new result—e.g., a mathematical theorem whose proof, or even whose statement, no human can understand. That is philosophically different from the way we have been doing science. Or at least thought we had; some might argue that we don't know how our own brains reach conclusions either, and that these new methods are a way of mimicking learning by the human brain. Nevertheless, I find this potential loss of understanding disturbing.

Despite the remarkable advances in computing, the hype about AGI—a general-intelligence machine that will think like a human and possibly develop consciousness—smacks of science fiction to me, partly because we don't understand the brain at that level of detail. Not only do we not understand what consciousness is, we don't even understand a relatively simple problem like how we remember a phone number. In just that one question, there are all sorts of things to consider. How do we know it is a number? How do we associate it with a person, a name, face, and other characteristics? Even such seemingly trivial questions involve everything from high-level cognition and memory to how a cell stores information and how neurons interact.

Moreover, that's just one task among many that the brain does effortlessly. Whereas machines will no doubt do ever more amazing things, they're unlikely to be a replacement for human thought and human creativity and vision. Eric Schmidt, former chairman of Google's parent company, said in a recent interview at the London Science Museum that even designing a robot that would clear the table, wash the dishes, and put them away was a huge challenge. The calculations involved in figuring out all the movements the body has to make to throw a ball accurately or do slalom skiing are prodigious. The brain can do all these and also do mathematics and music, and *invent* games like chess and Go, not just play them. We tend to underestimate the complexity and creativity of the human brain and how amazingly general it is.

If AI is to become more humanlike in its abilities, the machine-learning and neuroscience communities need to interact closely, something that is happening already. Some of today's greatest exponents of machine learning—such as Geoffrey Hinton, Zoubin Ghahramani, and Demis Hassabis—have backgrounds in cognitive neuroscience, and their success has been at least in part due to attempts to model brainlike behavior in their algorithms. At the same time, neurobiology has also flourished. All sorts of tools have been developed to watch which neurons are firing and genetically manipulate them and see what's happening in real time with inputs. Several countries have launched moon-shot neuroscience initiatives to see if we can crack the workings of the brain.

Advances in AI and neuroscience seem to go hand in hand; each field can propel the other.

Many evolutionary scientists, and such philosophers as Daniel Dennett, have pointed out that the human brain is the result of billions of years of evolution.* Human intelligence is not the special characteristic we think it is, but just another survival mechanism not unlike our digestive and immune systems, both of which are also amazingly complex. Intelligence evolved because it allowed us to make sense of the world around us, to plan ahead, and thus to cope with all sorts of unexpected things in order to survive. However, as Descartes stated, we humans define our very existence by our ability to think. So it is not surprising that, in an anthropomorphic way, our fears about AI reflect this belief that our intelligence is what makes us special.

But if we step back and look at life on Earth, we see that we are far from the most resilient species. If we're going to be taken over at some point, it will be by some of Earth's oldest life-forms, like bacteria, which can live anywhere from Antarctica to deep-sea thermal vents hotter than boiling water, or in acid environments that would melt you and me. So when people ask where we're headed, we need to put the question in a broader context. I don't know what sort of future AI will bring: whether AI will make humans subservient or obsolete or will be a useful and welcome enhancement of our abilities that will enrich our lives. But I am reasonably certain that computers will never be the overlords of bacteria.

---

* See, for example, Dennett's *From Bacteria to Bach and Back: The Evolution of Minds* (New York: W. W. Norton, 2017).

Chapter 19

THE HUMAN STRATEGY

ALEX "SANDY" PENTLAND

Alex "Sandy" Pentland is Toshiba Professor and professor of media arts and sciences at MIT; director of the Human Dynamics and Connection Science labs and the Media Lab Entrepreneurship Program; and the author of Social Physics.

Alex "Sandy" Pentland, an exponent of what he has termed "social physics," is interested in building powerful human-AI ecologies. He is concerned at the same time about the potential dangers of decision-making systems in which the data in effect take over and human creativity is relegated to the background.

The advent of Big Data, he believes, has given us the opportunity to reinvent our civilization: "We can now begin to actually look at the details of social interaction and how those play out, and we're no longer limited to averages like market indices or election results. This is an astounding change. The ability to see the details of the market, of political revolutions, and to be able to predict and control them is definitely a case of Promethean fire—it could be used for good or for ill. Big Data brings us to interesting times."

At our group meeting in Washington, Connecticut, he confessed that reading Norbert Wiener on the concept of feedback "felt like reading my own thoughts."

"After Wiener, people discovered or focused on the fact that there are genuinely chaotic systems that are just not predictable," he said, "but if you look at human socioeconomic systems, there is a large percentage of variance you can account for and predict. . . . Today there is data from all sorts of digital devices, and from all of our transactions. The fact that everything is datafied means you can measure things in real time in most aspects of human life—and increasingly in every aspect of human life. The fact that we have interesting computers and machine-learning techniques means that you can build predictive models of human systems in ways you could never do before."

n the last half century, the idea of AI and intelligent robots has dominated thinking about the relationship between humans and computers. In part, this is because it's easy to tell the stories about AI and robots, and in part because of early successes (e.g., theorem provers that reproduced most of Whitehead and Russell's *Principia Mathematica*) and massive military funding. The earlier and broader vision of cybernetics, which considered the artificial as part of larger systems of feedback and mutual influence, faded from public awareness.

However, in the intervening years the cybernetics vision has slowly grown and quietly taken over—to the point where it is "in the air." State-of-the-art research in most engineering disciplines is now framed as feedback systems that are dynamic and driven by energy flows. Even AI is being recast as human/machine "adviser" systems, and the military is beginning large-scale funding in this area—something that should perhaps worry us more than drones and independent humanoid robots.

But as science and engineering have adopted a more cybernetics-like stance, it has become clear that even the vision of cybernetics is far too small. It was originally centered on the embeddedness of the individual actor but not on the emergent properties of a network of actors. This is unsurprising, because the mathematics of networks did not exist until recently, so a quantitative science of how networks behave was impossible. We now know that study of the individual does not produce understanding of the system except in certain simple cases. Recent progress in this area was foreshadowed by understanding that "chaos" and, later, "complexity" were the typical behavior of systems, but we can now go far beyond these statistical understandings.

We're beginning to be able to analyze, predict, and even design the emergent behavior of complex heterogeneous networks. The cybernetics view of the connected individual actor can now be expanded to cover complex systems of connected individuals and machines, and the insights we obtain from this broader view are fundamentally different from those obtained from the cybernetics view. Thinking about the network is analogous to thinking about entire ecosystems. How would you guide ecosystems to grow in a good direction? What do you even mean by "a good direction"? Questions like these are beyond the boundary of traditional cybernetic thinking.

Perhaps the most stunning realization is that humans are already beginning to use AI and machine learning to guide entire ecosystems, including ecosystems of people, thus creating human-AI ecologies. Now that everything is becoming "datafied," we can measure most aspects of human life and, increasingly, aspects of all life. This, together with new, powerful machine-learning techniques, means that we can build models of these ecologies in ways we couldn't before. Well-known examples are weather- and traffic-prediction models, which are being extended to predict the global climate and plan city growth and renewal. AI-aided engineering of the ecologies is already here.

Development of human-AI ecosystems is perhaps inevitable for a social species such as ourselves. We became social early in our evolution, millions of years ago. We began exchanging information with one another to stay alive, to increase our fitness. We developed writing to share abstract and complex ideas, and most recently we've developed computers to enhance our communication abilities. Now we're developing AI and machine-learning models of ecosystems and sharing the predictions of those models to jointly shape our world through new laws and international agreements.

We live in an unprecedented historic moment, in which the availability of vast amounts of human behavioral data and advances in machine learning enable us to tackle complex social problems through algorithmic decision making. The opportunities for such a human-AI ecology to have positive social impact through fairer and more transparent decisions are obvious. But there are also risks of a "tyranny of algorithms," where unelected data experts are running the world. The choices we make now are perhaps even more momentous than those we faced in the 1950s, when AI and cybernetics were created. The issues look similar, but they're not. We have moved down the road, and now the scope is larger. It's not just AI robots versus individuals. It's AI guiding entire ecologies.

How can we make a good human-artificial ecosystem, something that's not a machine society but a cyberculture in which we can all live as humans—a culture with a human feel to it? We don't want to think small—for example, to talk only of robots and self-driving cars. We want this to be a global ecology. Think Skynet size. But how would you make Skynet something that's about the human fabric?

The first thing to ask is: What's the magic that makes the current AI work? Where is it wrong and where is it right?

The good magic is that it has something called the credit-assignment function. What that lets you do is take "stupid neurons"—little linear functions—and figure out, in a big network, which ones are doing the work and strengthen them. It's a way of taking a random bunch of switches all hooked together in a network and making them smart by giving them feedback about what works and what doesn't. This sounds simple, but there's some complicated math around it. That's the magic that makes current AI work.

The bad part of it is that because those little neurons are stupid, the things they learn don't generalize very well. If an AI sees something it hasn't seen before, or if the world changes a little bit, the AI is likely to make a horrible mistake. It has absolutely no sense of context. In some ways, it's as far from Norbert Wiener's original notion of cybernetics as you can get, because it isn't contextualized; it's a little idiot savant.

But imagine that you took away those limitations: Imagine that instead of using dumb neurons, you used neurons in which real-world knowledge was embedded. Maybe instead of linear neurons, you used neurons that were functions in physics, and then you tried to fit physics data. Or maybe you put in a lot of knowledge about humans and how they interact with one another—the statistics and characteristics of humans.

When you add this background knowledge and surround it with a good credit-assignment function, then you can take observational data and use the credit-assignment function to reinforce the functions that are producing good answers. The result is an AI that works extremely well and can generalize. For instance, in solving physical problems, it often takes only a couple of noisy data points to get something that's a beautiful description of a phenomenon, because you're putting in knowledge about how physics works. That's in huge contrast to normal AI, which requires millions of training examples and is very sensitive to noise. By adding the appropriate background knowledge you get much more intelligence.

Similar to the physical-systems case, if we make neurons that know a lot about how humans learn from one another, then we can detect human fads and predict human behavior trends in surprisingly accurate and efficient ways. This "social physics" works because human behavior is determined as much by the patterns of our culture as by rational, individual thinking. These patterns can be described mathematically and employed to make accurate predictions.

This idea of a credit-assignment function reinforcing connections between neurons that are doing the best work is the core of current AI. If you make those little neurons smarter, the AI gets smarter. So what would happen if we replaced the neurons with people? People have lots of capabilities. They know lots of things about the world; they can perceive things in a broadly competent, human way. What would happen if you had a network of people in which you could reinforce the connections that were helping and minimize the connections that weren't?

That begins to sound like a society, or a company. We all live in a human social network. We're reinforced for doing things that seem to help everybody and discouraged from doing things that are not appreciated. Culture is the result of this sort of human AI as applied to human problems; it is the process of building social structures by reinforcing the good connections and penalizing the bad. Once you've realized you can take this general AI framework and create a human AI, the question becomes, What's the right way to do that? Is it a safe idea? Is it completely crazy?

My students and I are looking at how people make decisions, on huge databases of financial decisions, business decisions, and many other sorts of decisions. What we've found is that humans often make decisions in a way that mimics AI credit-assignment algorithms and works to make the community smarter. A particularly interesting feature of this work is that it addresses a classic problem in evolution known as the group-selection problem. The core of this problem is: How can we select for culture in evolution, when it's the individuals that reproduce? What you need is something that selects for the best cultures and the best groups but also selects for the best individuals, because they're the units that transmit the genes.

When you frame the question this way and go through the mathematical literature, you discover that there's one generally best way to do this. It's called "distributed Thompson sampling," a mathematical algorithm used in choosing, out of a set of possible actions with unknown payoffs, the action that maximizes the expected reward in respect to the actions. The key is social sampling, a way of combining evidence, of exploring and exploiting at the same time. It has the unusual property of simultaneously being the best strategy both for the individual and for the group. If you use the group as the basis of selection, and then the group gets either wiped out or reinforced, you're also selecting for successful individuals. If you select for individuals, and each individual does what's good for him or her, then that's automatically the best thing for the group. It's an amazing alignment of interests and utilities, and it provides real insight into the question of how culture fits into natural selection.

Social sampling, very simply, is looking around you at the actions of people who are like you, finding what's popular, and then copying it if it seems like a good idea to you. Idea propagation has this popularity function driving it, but individual adoption also is about figuring out how the idea works for the individual—a reflective attitude. When you combine social sampling and personal judgment, you get superior decision making. That's amazing, because now we have a mathematical recipe for doing with humans what all those AI techniques are doing with dumb computer neurons. We have a way of putting people together to make better decisions, given more and more experience.

So what happens in the real world? Why don't we do this all the time? Well, people are good at it, but there are ways it can run amok. One of these is through advertising, propaganda, or "fake news." There are many ways to get people to think something is popular when it's not, and this destroys the usefulness of social sampling. The way you can make groups of people smarter, the way you can make human AI, will work only if you can get feedback to them that's truthful. It must be grounded on whether each person's actions worked for them or not.

That's the key to AI mechanisms, too. What they do is analyze whether they performed correctly. If so, plus one; if not, minus one. We need that truthful feedback to make this human mechanism work well, and we need good ways of knowing about what other people are doing so that we can correctly assess popularity and the likelihood of this being a good choice.

The next step is to build this credit-assignment function, this feedback function, for people, so that we can make a good human-artificial ecosystem—a smart organization and a smart culture. In a way, we need to duplicate some of the early insights that resulted in, for instance, the U.S. census—trying to find basic facts that everybody can agree on and understand so that the transmission of knowledge and culture can happen in a way that's truthful and social sampling can function efficiently.

We can address the problem of building an accurate credit-assignment function in many different settings. In companies, for instance, it can be done with digital ID badges that reveal who's connected to whom, so that we can assess the pattern of connections in relation to the company's results on a daily or weekly basis. The credit-assignment function asks whether those connections helped solve problems, or helped invent new solutions, and reinforces the helpful connections. When you can get that feedback quantitatively—which is difficult, because most things aren't measured quantitatively—both the productivity and the innovation rate within the organization can be significantly improved. This is, for instance, the basis of Toyota's "continuous improvement" method.

A next step is to try to do the same thing but at scale, something I refer to as building a trust network for data. It can be thought of as a distributed system like the Internet, but with the ability to quantitatively measure and communicate the qualities of human society, in the same way that the U.S. census does a pretty good job of telling us about population and life expectancy. We are already deploying prototype examples of trust networks at scale in several countries, based on the data and measurement standards laid out in the U.N. Sustainable Development Goals.

On the horizon is a vision of how we can make humanity more intelligent by building a human AI. It's a vision composed of two threads. One is data that we can all trust—data that have been vetted by a broad community, data where the algorithms are known and monitored, much like the census data we all automatically rely on as at least approximately correct. The other is a fair, data-driven assessment of public norms, policy, and government, based on trusted data about current conditions. This second thread depends on availability of trusted data and so is just beginning to be developed. Trusted data and data-driven assessment of norms, policy, and government together create a credit-assignment function that improves societies' overall fitness and intelligence.

It is precisely at the point of creating greater societal intelligence where fake news, propaganda, and advertising all get in the way. Fortunately, trust networks give us a path forward to building a society more resistant to echo-chamber problems, these fads, these exercises in madness. We have begun to develop a new way of establishing social measurements in aid of curing some of the ills we see in society today. We're using open data from all sources, encouraging a fair representation of the things people are choosing, in a curated mathematical framework that can stamp out the echoes and the attempts to manipulate us

ON POLARIZATION AND INEQUALITY

Extreme polarization and segregation by income are almost everywhere in the world today and threaten to tear governments and civil society apart. Increasingly, the media are becoming adrenaline pushers driven by advertising clicks and failing to deliver balanced facts and reasoned discourse—and the degradation of media is causing people to lose their bearings. They don't know what to believe, and thus they can easily be manipulated. There is a real need to ground our various cultures in trustworthy, data-driven standards that we all agree on, and to be able to know what behaviors and policies work and which don't.

In converting to a digital society, we've lost touch with traditional notions of truth and justice. Justice used to be mostly informal and normative. We've now formalized it. At the same time, we've put it out of reach for most people. Our legal systems are failing us in a way they didn't before, precisely because they're now more formal, more digital, less embedded in society.

Ideas about justice are very different around the world. One of the core differentiators is this: Do you or your parents remember when the bad guys came with guns and took everything? If you do, your attitude about justice is different from that of the average reader of this essay. Do you come from the upper classes? Or were you somebody who saw the sewers from the inside? Your view of justice depends on your history.

A common test I have for U.S. citizens is this: Do you know anybody who owns a pickup truck? It's the number-one-selling vehicle in the United States, and if you don't know people like that, you're out of touch with more than 50 percent of Americans. Physical segregation drives conceptual segregation. Most of America thinks of justice and access and fairness in terms very different from those of the typical, say, Manhattanite.

If you look at patterns of mobility—where people go—in a typical city, you find that the people in the top quintile (white-collar working families) and bottom quintile (people who are sometimes on unemployment or welfare) almost never talk to one another. They don't go to the same places; they don't talk about the same things. They all live in the same city, nominally, but it's as if it were two completely different cities—and this is perhaps the most important cause of today's plague of polarization.

ON EXTREME WEALTH

Some two hundred of the world's wealthiest people have pledged to give away more than 50 percent of their wealth either during their lifetimes or in their wills, creating a plurality of voices in the foundation space.* Bill Gates is probably the most familiar example. He's decided that if the government won't do it, he'll do it. You want mosquito nets? He'll do it. You want antivirals? He'll do it. We're getting different stakeholders to take action in the form of foundations dedicated to public good, and they have different versions of what they consider the public good. This diversity of goals has created a lot of what's wonderful about the world today. Actions from outside government by organizations like the Ford Foundation and the Sloan Foundation, that bet on things that nobody else would bet on, have changed the world for the better.

Sure, these billionaires are human, with human foibles, and all is not

---

* https://givingpledge.org/About.aspx.

necessarily as it should be. On the other hand, the same situation obtained when the railways were first built. Some people made huge fortunes. A lot of people went bust. We, the average people, got railways out of it. That's good. Same thing with electric power; same thing with many new technologies. There's a churning process that throws somebody up and later casts them or their heirs down. Bubbles of extreme wealth were a feature of the late 1800s and early 1900s when steam engines and railways and electric lights were invented. The fortunes they created were all gone within two or three generations.

If the U.S. were like Europe, I would worry. What you find in Europe is that the same families have held on to wealth for hundreds of years, so they're entrenched in terms not just of wealth but of the political system and in other ways. But so far, the U.S. has avoided this kind of hereditary class system. Extreme wealth hasn't stuck, which is good. It shouldn't stick. If you win the lottery, you get your billion dollars, but your grandkids ought to work for a living.

ON AI AND SOCIETY

People are scared about AI. Perhaps they should be. But they need to realize that AI feeds on data. Without data, AI is nothing. You don't have to watch the AI; instead you should watch what it eats and what it does. The trust-network framework we've set up, with the help of nations in the EU and elsewhere, is one where we can have our algorithms, we can have our AI, but we get to see what went in and what went out, so that we can ask, Is this a discriminatory decision? Is this the sort of thing that we want as humans? Or is this something that's a little weird?

The most revealing analogy is that regulators, bureaucracies, and parts of the government are very much like AIs: They take in the rules that we call law and regulation, and they add government data, and they make decisions that affect our lives. The part that's bad about the current system is that we have very little oversight of these departments, regulators, and bureaucracies. The only control we have is the vote—the opportunity to elect somebody different. We need to make oversight of bureaucracies a lot more fine-grained. We need to record the data that went into every single decision and have the results analyzed by the various stakeholders—rather like elected legislatures were originally intended to do.

If we have the data that go into and out of each decision, we can easily ask, Is this a fair algorithm? Is this AI doing things that we as humans believe are ethical? This human-in-the-loop approach is called "open algorithms"; you get to see what the AIs take as input and what they decide using that input. If you see those two things, you'll know whether they're doing the right thing or the wrong thing. It turns out that's not hard to do. If you control the data, then you control the AI.

One thing people often fail to mention is that all the worries about AI are the same as the worries about today's government. For most parts of the government—the justice system, etc.—there's no reliable data about what they're doing and in what situation. How can you know whether the courts are fair or not if you don't know the inputs and the outputs? The same problem arises with AI systems and is addressable in the same way. We need trusted data to hold current government to account in terms of what they take in and what they put out, and AI should be no different.

NEXT-GENERATION AI

Current AI machine-learning algorithms are, at their core, dead simple stupid. They work, but they work by brute force, so they need hundreds of millions of samples. They work because you can approximate anything with lots of little simple pieces. That's a key insight of current AI research—that if you use reinforcement learning for credit-assignment feedback, you can get those little pieces to approximate whatever arbitrary function you want.

But using the wrong functions to make decisions means the AI's ability to make good decisions won't generalize. If we give the AI new, different inputs, it may make completely unreasonable decisions. Or if the situation changes, then you need to retrain it. There are amusing techniques to find the "null space" in these AI systems. These are inputs that the AI thinks are valid examples of what it was trained to recognize (e.g., faces, cats, etc.), but to a human they're crazy examples.

Current AI is doing descriptive statistics in a way that's not science and would be almost impossible to make into science. To build robust systems, we need to know the science behind data. The systems I view as next-generation AIs result from this science-based approach: If you're going to create an AI to deal with something physical, then you should build the laws of physics into it as your descriptive functions, in place of those stupid little neurons. For instance, we know that physics uses functions like polynomials, sine waves, and exponentials, so those should be your basis functions and not little linear neurons. By using those more appropriate basis functions, you need a lot less data, you can deal with a lot more noise, and you get much better results.

As in the physics example, if we want to build an AI to work with human behavior, then we need to build the statistical properties of human networks into machine-learning algorithms. When you replace the stupid neurons with ones that capture the basics of human behavior, then you can identify trends with very little data, and you can deal with huge levels of noise.

The fact that humans have a "commonsense" understanding that they bring to most problems suggests what I call the human strategy: Human society is a network just like the neural nets trained for deep learning, but the "neurons" in human society are a lot smarter. You and I have surprisingly general descriptive powers that we use for understanding a wide range of situations, and we can recognize which connections should be reinforced. That means we can shape our social networks to work much better and potentially beat all that machine-based AI at its own game.

Chapter 20

MAKING THE INVISIBLE VISIBLE: ART MEETS AI

HANS ULRICH OBRIST

Hans Ulrich Obrist is artistic director of the Serpentine Gallery, London, and the author of Ways of Curating *and* Lives of the Artists, Lives of the Architects.

"URGENT! URGENT!" the cc'd copy of an email screamed, one of a dozen emails that greeted me as I turned on my phone at the baggage carousel at Malpensa Airport after the long flight from JFK. "The great American visionary thinker John Brockman arrives this morning at Grand Hotel Milan. You MUST, repeat MUST, pay him a visit." It was signed HUO.

The prior evening, waiting in the lounge at JFK, I had had the bright idea to write to my friend and longtime collaborator, the London-based, peripatetic art curator **Hans Ulrich Obrist** (known to all as HUO), and ask if there was anyone in Milan I should know.

Once I was settled at the hotel, the phone began ringing and a procession of leading Italian artists, designers, and architects called to request a meeting, including Enzo Mari, the modernist artist and furniture designer; Alberto Garutti, whose aesthetic strategies have inspired a dialogue between contemporary art, spectator, and public space; and fashion designer Miuccia Prada, who "requests your presence for tea this afternoon at Prada headquarters." And thus, thanks to HUO, did the jet-lagged "great American visionary thinker" stumble and mumble his way through his first day in Milan in November 2011.

HUO is sui generis: He lives a twenty-four-hour day, sleeping (I guess) whenever, and employing full-time assistants who work eight-hour shifts and are available to him 24/7. Over a recent two-year period, he visited art venues in either China or India for forty weekends each year—departing London on Thursday evening, back at his desk on Monday. Last year, once again, *ArtReview* ranked him number one on their annual "Power 100" list.

Recently we collaborated on a panel during the "GUEST, GHOST, HOST: MACHINE!" Serpentine event that took place at London's new City Hall. We were joined by Venki Ramakrishnan, Jaan Tallinn, and Andrew Blake, research director of The Alan Turing Institute. The event was consistent with HUO's mission of bringing together art and science: "The curator is no longer understood simply as the person who fills a space with objects," he says, "but also as the person who brings different cultural spheres into contact, invents new display features, and makes junctions that allow unexpected encounters and results."

In the introduction to the second edition of his book *Understanding Media*, Marshall McLuhan noted the ability of art to "anticipate future social and technological developments." Art is "an early alarm system," pointing us to new developments in times ahead and allowing us "to prepare to cope with them. . . . Art as a radar environment takes on the function of indispensable perceptual training."

In 1964, when McLuhan's book was first published, the artist Nam June Paik was just building his *Robot K-456* to experiment with the technologies that subsequently would start to influence society. He had worked with television earlier, challenging its usual passive consumption by the viewer, and later made art with global live-satellite broadcasts, using the new media less for entertainment than to point us to their poetic and intercultural capacities (which are still mostly unused today). The Paiks of our time, of course, are now working with the Internet, digital images, and artificial intelligence. Their works and thoughts, again, are an early alarm system for the developments ahead of us.

As a curator, my daily work is to bring together different works of art and connect different cultures. Since the early 1990s, I have also been organizing conversations and meetings with practitioners from different disciplines, in order to go beyond the general reluctance to pool knowledge. Since I was interested in hearing what artists have to say about artificial intelligence, I recently organized several conversations between artists and engineers.

The reason to look closely at AI is that two of the most important questions of today are "How capable will AI become?" and "What dangers may arise from it?" Its early applications already influence our everyday lives in ways that are more or less recognizable. There is an increasing impact on many aspects of our society, but whether this might be, in general, beneficial or malign is still uncertain.

Many contemporary artists are following these developments closely. They are articulating various doubts about the promises of AI and reminding us not to associate the term "artificial intelligence" solely with positive outcomes. To the current discussions of AI, the artists contribute their specific perspectives and notably their focus on questions of image making, creativity, and the use of programming as artistic tools.

The deep connections between science and art had already been noted by the late Heinz von Foerster, one of the architects of cybernetics, who worked with Norbert Wiener from the mid-1940s and in the 1960s founded the field of second-order cybernetics, in which the observer is understood as part of the system itself and not as an external entity. I knew von Foerster well, and in one of our many conversations he offered his views on the relation between art and science:

> I've always perceived art and science as complementary fields. One shouldn't forget that a scientist is in some respects also an artist. He invents a new technique and he describes it. He uses language like a poet, or the author of a detective novel, and describes his findings. In my view, a scientist must work in an artistic way if he wants to communicate his research. He obviously wants to communicate and talk to others. A scientist invents new objects, and the question is how to describe them. In all of these aspects, science is not very different from art.

When I asked him how he defined cybernetics, von Foerster answered:

The substance of what we have learned from cybernetics is to think in circles: *A* leads to *B, B* to *C,* but *C* can return to *A.* Such kinds of arguments are not linear but circular. The significant contribution of cybernetics to our thinking is to accept circular arguments. This means that we have to look at circular processes and understand under which circumstances an equilibrium, and thus a stable structure, emerges.

Today, where AI algorithms are applied in daily tasks, one can ask how the human factor is included in these kinds of processes and what role creativity and art could play in relation to them. There are thus different levels to think about when exploring the relation between AI and art.

So what do contemporary artists have to say about artificial intelligence?

ARTIFICIAL STUPIDITY

Hito Steyerl, an artist who works with documentary and experimental film, considers two key aspects that we should keep in mind when reflecting on the implications of AI for society. First, the expectations for so-called artificial intelligence, she says, are often overrated, and the noun "intelligence" is misleading; to counter that, she uses the term "artificial stupidity." Second, she points out that programmers are now making invisible software algorithms visible through images, but to understand and interpret these images better, we should apply the expertise of artists.

Steyerl has worked with computer technology for many years, and her recent artworks have explored surveillance techniques, robots, and such computer games as in *How Not to be Seen* (2013), on digital-image technologies, or *HellYeahWeFuckDie* (2017), about the training of robots in the still-difficult task of keeping balance. But to explain her notion of artificial stupidity, Steyerl refers to a more general phenomenon, like the now widespread use of Twitter bots, noting in our conversation:

It was and still is a very popular tool in elections to deploy Twitter armies to sway public opinion and deflect popular hashtags and so on. This is an artificial intelligence of a very, very low grade. It's two or maybe three lines of script. It's nothing very sophisticated at all. Yet the social implications of this kind of artificial stupidity, as I call it, are already monumental in global politics.

As has been widely noted, this kind of technology was seen in the many automated Twitter posts before the 2016 U.S. presidential election and also shortly before the Brexit vote. If even low-grade AI technology like these bots are already influencing our politics, this raises another urgent question: How powerful will far more advanced techniques be in the future?

VISIBLE/INVISIBLE

The artist Paul Klee often talked about art as "making the invisible visible." In computer technology, most algorithms work invisibly, in the background; they remain inaccessible in the systems we use daily. But lately there has been an interesting comeback of visuality in machine learning. The ways that the deep-learning algorithms of AI are processing data have been made visible through applications like Google's DeepDream, in which the process of computerized pattern recognition is visualized in real time. The application shows how the algorithm tries to match animal forms with any given input. There are many other AI visualization programs that, in their way, also "make the invisible visible." The difficulty in the general public perception of such images is, in Steyerl's view, that these visual patterns are viewed uncritically as realistic and objective representations of the machine process. She says of the aesthetics of such visualizations:

> For me, this proves that science has become a subgenre of art history. . . . We now have lots of abstract computer patterns that might

look like a Paul Klee painting, or a Mark Rothko, or all sorts of other abstractions that we know from art history. The only difference, I think, is that in current scientific thought they're perceived as representations of reality, almost like documentary images, whereas in art history there's a very nuanced understanding of different kinds of abstraction.

What she seeks is a more profound understanding of computer-generated images and the different aesthetic forms they use. They are obviously not generated with the explicit goal of following a certain aesthetic tradition. The computer engineer Mike Tyka, in a conversation with Steyerl, explained the functions of these images:

> Deep-learning systems, especially the visual ones, are really inspired by the need to know what's going on in the black box. Their goal is to project these processes back into the real world.

Nevertheless, these images have aesthetic implications and values that have to be taken into account. One could say that while the programmers use these images to help us better understand the programs' algorithms, we need the knowledge of artists to better understand the aesthetic forms of AI. As Steyerl has pointed out, such visualizations are generally understood as "true" representations of processes, but we should pay attention to their respective aesthetics, and their implications, which have to be viewed in a critical and analytical way.

In 2017, the artist Trevor Paglen created a project to make these invisible AI algorithms visible. In *Sight Machine*, he filmed a live performance of the Kronos Quartet and processed the resulting images with various computer software programs used for face detection, object identification, and even missile guidance. He projected the outcome of these algorithms, in real time, back to screens above the stage. By demonstrating how the various different programs interpreted the musicians' performance, Paglen showed that AI algorithms are always determined by sets of values and interests that they then manifest and reiterate, and thus must be critically questioned. The significant contrast between algorithms and music also raises the issue of relationships between technical and human perception.

COMPUTERS, AS A TOOL FOR CREATIVITY, CAN'T REPLACE THE ARTIST

Rachel Rose, a video artist who thinks about the questions posed by AI, employs computer technology in the creation of her works. Her films give the viewer an experience of materiality through the moving image. She uses collaging and layering of the material to manipulate sound and image, and the editing process is perhaps the most important aspect of her work.

She also talks about the importance of decision making in her work. For her, the artistic process does not follow a rational pattern. In a conversation we had, together with the engineer Kenric McDowell, at the Google Cultural Institute, she explained this by citing a story from theater director Peter Brook's 1968 book *The Empty Space*. When Brook designed the set for his production of *The Tempest* in the late 1960s, he started by making a Japanese garden, but then the design evolved, becoming a white box, a black box, a realistic set, and so on. And in the end, he returned to his original idea. Brook writes that he was shocked at having spent a month on his labors, only to end at the beginning. But this shows that the creative artistic process is a succession whose every step builds on the next and which eventually comes to an unpredictable conclusion. The process is not a logical or rational succession but has mostly to do with the artist's feelings in reaction to the preceding result. Rose said, of her own artistic decision making:

> It, to me, is distinctively different from machine learning, because at each decision there's this core feeling that comes from a human being, which has to do with empathy, which has to do with communication, which has to do with questions about our own mortality that only a human could ask.

This point underlines the fundamental difference between any human artistic production and so-called computer creativity. Rose sees AI more as a possible way to create better tools for humans:

> A place I can imagine machine learning working for an artist would be not in developing an independent subjectivity, like writing a poem or making an image, but actually in filling in gaps that are to do with labor, like the way that Photoshop works with different tools that you can use.

And though such tools may not seem spectacular, she says, "they might have a larger influence on art," because they provide artists with further possibilities in their creative work.

McDowell added that he, too, believes there are false expectations around AI. "I've observed," he said, "that there's a sort of magical quality to the idea of a computer that does all the things that we do." He continued: "There's almost this kind of demonic mirror that we look into, and we want it to write a novel, we want it to make a film—we want to give that away somehow." He is instead working on projects wherein humans collaborate with the machine. One of the current aims of AI research is to find new means of interaction between humans and software. And art, one could say, needs to play a key role in that enterprise, since it focuses on our subjectivity and on essential human aspects like empathy and mortality.

CYBERNETICS/ART

Suzanne Treister is an artist whose work from 2009 to 2011 serves as an example of what is happening at the intersection of our current technologies, the arts, and cybernetics. Treister has been a pioneer in digital art since the 1990s, inventing, for example, imaginary video games and painting screen shots from them. In her project *HEXEN 2.0* she looked back at the famous Macy Conferences on cybernetics that between 1946 and 1953 were organized in New York by engineers and social scientists to unite the sciences and to develop a universal theory of the workings of the mind.

In her project, she created thirty photo-text works about the conference attendees (which included Wiener and von Foerster), she invented tarot cards, and she made a video based on a photomontage of a "cybernetic séance." In the "séance," the conference participants are seen sitting at a round table, as in spiritualist séances, while certain of their statements on cybernetics are heard in an audio collage—rational knowledge and superstition combined. She also noted that some of the participating scientists worked for the military; thus the application of cybernetics could be seen in an ambivalent way, even back then, as a tussle between pure knowledge and its use in state control.

If one looks at Treister's work about the Macy Conferences participants, one sees that no visual artist was included. A dialogue between artists and scientists would be fruitful in future discussions, and it is a bit astonishing that this wasn't realized at the time, given von Foerster's keen interest in art. He recounted in one of our conversations how his relation to the field dated back to his childhood:

I grew up as a child in an artistic family. We often had visits from poets, philosophers, painters, and sculptors. Art was a part of my life. Later, I got into physics, as I was talented in this subject. But I always remained conscious of the importance of art for science. There wasn't a great difference for me. For me, both aspects of life have always been very much alike—and accessible, too. We should see them as one. An artist also has to reflect on his work. He has to think about his grammar and his language. A painter must know how to handle his colors. Just think of how intensively oil colors were researched during the Renaissance. They wanted to know how a certain pigment could be mixed with others to get a certain tone of red or blue. Chemists and painters collaborated very closely. I think the artificial division between science and art is wrong.

Though for von Foerster the relation between art and science was always clear, for our own time this connection remains to be made. There are many reasons to multiply the links. The critical thinking of artists would be beneficial in respect to the dangers of AI, since they draw our attention to questions they consider essential from their perspective. With the advent of machine learning, new tools are available to artists for their work. And as the algorithms of AI are made visible through artificial images in new ways, artists' critical visual knowledge and expertise will be harnessed. Many of the key questions of AI are philosophical in nature and can be answered only from a holistic point of view. The way they play out among adventurous artists will be worth following.

SIMULATING WORLDS

For the most part, the works of contemporary artists have been embodied ruminations on AI's impact on existential questions of the self and our future interaction with nonhuman entities. Few, though, have taken the technologies and innovations of AI as the underlying materials of their work and sculpted them to their own vision. An exception is the artist Ian Cheng, who has gone as far as to construct entire worlds of artificial beings with varying degrees of sentience and intelligence. He refers to these worlds as Live Simulations. His *Emissaries* trilogy (2015–17) is set in a fictional postapocalyptic world of flora and fauna, in which AI-driven animals and creatures explore the landscape and interact with one another. Cheng uses advanced graphics but has them programmed with a lot of glitches and imperfections, which imparts a futuristic and anachronistic atmosphere at the same time. Through his trilogy, which charts a history of consciousness, he asks the question "What is a simulation?"

While the majority of artistic works that utilize recent developments in AI specifically draw from the field of machine learning, Cheng's Live

Simulations take a separate route. The protagonists and plotlines that are interlaced in each episodic simulation of *Emissaries* use the complex logic systems and rules of AI. What is profound about his continually evolving scenes is that complexity arises not through the desires/actions of any single actor or artificial godhead but instead through their constellation, collision, and constant evolution in symbiosis with one another. This gives rise to unexpected outcomes and unending, unknowable situations—you can never experience the exact same moment in successive viewings of his work.

Cheng had a discussion at the Serpentine Marathon "GUEST, GHOST, HOST: MACHINE!" with the programmer Richard Evans, who recently designed Versu, an AI-based platform for interactive storytelling games. Evans's work emphasizes the social interaction of the games' characters, who react in a spectrum of possible behaviors to the choices made by the human players. In their conversation, Evans said that a starting point for the project was that most earlier simulation video games, such as *The Sims*, did not sufficiently take into account the importance of social practices. Simulated protagonists in games would often act in ways that did not correspond well with real human behavior. Knowledge of social practices limits the possibilities of action but is necessary to understand the meaning of our actions—which is what interests Cheng for his own simulations. The more parameters of actions in certain circumstances are determined in a computer simulation, the more interesting it is for Cheng to experiment with individual and specific changes. He told Evans, "I gather that if we had AI with more ability to respond to social contexts, tweaking one thing, you would get something quite artistic and beautiful."

Cheng also sees the work of programmers and AI simulations as creating new and sophisticated tools for experimenting with the parameters of our daily social practices. In this way, the involvement of artists in AI will lead to new kinds of open experiments in art. Such possibilities are—like increased AI capabilities in general—still in the future. Recognizing that this is an experimental technology in its infancy, very far from apocalyptic visions of a superintelligent AI takeover, Cheng fills his simulations with prosaic avatars such as strange microbial globules, dogs, and the undead.

Discussions like these, between artists and engineers, of course are not totally new. In the 1960s, the engineer Billy Klüver brought artists together with engineers in a series of events, and in 1967 he founded the Experiments in Art and Technology program with Robert Rauschenberg and others. In London at around the same time, Barbara Steveni and John Latham, of the Artist Placement Group, took things a step further by asserting that there should be artists in residence in every company and every government. Today, these inspiring historical models can be applied to the field of AI. As AI comes to inhabit more and more of our everyday lives, the creation of a space that is nondeterministic and nonutilitarian in its plurality of perspectives and diversity of understandings will undoubtedly be essential.

Chapter 21

AIs VERSUS FOUR-YEAR-OLDS

ALISON GOPNIK

Alison Gopnik is a developmental psychologist at UC Berkeley. Her books include The Philosophical Baby *and, most recently,* The Gardener and the Carpenter: What the New Science of Child Development Tells Us About the Relationship Between Parents and Children.

Alison Gopnik is an international leader in the field of children's learning and development and was one of the founders of the field of "theory of mind." She has spoken of the child brain as a "powerful learning computer," perhaps from personal experience. Her own Philadelphia childhood was an exercise in intellectual development. "Other families took their kids to see *The Sound of Music* or *Carousel;* we saw Racine's *Phaedra* and Samuel Beckett's *Endgame*," she has recalled. "Our family read Henry Fielding's eighteenth-century novel *Joseph Andrews* out loud to each other around the fire on camping trips."

Lately she has invoked Bayesian models of machine learning to explain the remarkable ability of preschoolers to draw conclusions about the world around them without benefit of enormous data sets. "I think babies and children are actually more conscious than we are as adults," she has said. "They're very good at taking in lots of information from lots of different sources at once." She has referred to babies and young children as "the research and development division of the human species." Not that she treats them coldly, as if they were mere laboratory animals. They appear to revel in her company, and in the blinking, thrumming toys in her Berkeley lab. For years after her own children had outgrown it, she kept a playpen in her office.

Her investigations into just how we learn, and the parallels to the deep-learning methods of AI, continue. "It turns out to be much easier to simulate the reasoning of a highly trained adult expert than to mimic the ordinary learning of every baby," she says. "Computation is still the best—indeed, the only—scientific explanation we have of how a physical object like a brain can act intelligently. But, at least for now, we have almost no idea at all how the sort of creativity we see in children is possible."

Everyone's heard about the new advances in artificial intelligence, and especially machine learning. You've also heard utopian or apocalyptic predictions about what those advances mean. They have been taken to presage either immortality or the end of the world, and a lot has been written about both of those possibilities. But the most sophisticated AIs are still far from being able to solve problems that human four-year-olds accomplish with ease. In spite of the impressive name, artificial intelligence largely consists of techniques to detect statistical patterns in large data sets. There is much more to human learning.

How can we possibly know so much about the world around us? We learn an enormous amount even when we are small children; four-year-olds already know about plants and animals and machines; desires, beliefs, and emotions; even dinosaurs and spaceships.

Science has extended our knowledge about the world to the unimaginably large and the infinitesimally small, to the edge of the universe and the beginning of time. And we use that knowledge to make new classifications and predictions, imagine new possibilities, and make new things happen in the world. But all that reaches any of us from the world is a stream of photons hitting our retinas and disturbances of air at our eardrums. How do we learn so much about the world when the evidence we have is so limited? And how do we do all this with the few pounds of grey goo that sits behind our eyes?

The best answer so far is that our brains perform computations on the concrete, particular, messy data arriving at our senses, and those computations yield accurate representations of the world. The representations seem to be structured, abstract, and hierarchical; they include the perception of three-dimensional objects, the grammars that underlie language, and mental capacities like "theory of mind," which lets us understand what other people think. Those representations allow us to make a wide range of new predictions and imagine many new possibilities in a distinctively creative human way.

This kind of learning isn't the only kind of intelligence, but it's a particularly important one for human beings. And it's the kind of intelligence that is a specialty of young children. Although children are dramatically bad at planning and decision making, they are the best learners in the universe. Much of the process of turning data into theories happens before we are five.

Since Aristotle and Plato, there have been two basic ways of addressing the problem of how we know what we know, and they are still the main approaches in machine learning. Aristotle approached the problem from the bottom up: Start with senses—the stream of photons and air vibrations (or the pixels or sound samples of a digital image or recording)—and see if you can extract patterns from them. This approach was carried further by such classic associationists as philosophers David Hume and J. S. Mill and later by behavioral psychologists, like Pavlov and B. F. Skinner. On this view, the abstractness and hierarchical structure of representations is something of an illusion, or at least an epiphenomenon. All the work can be done by association and pattern detection—especially if there are enough data.

Over time, there has been a seesaw between this bottom-up approach to the mystery of learning and Plato's alternative, top-down one. Maybe we get abstract knowledge from concrete data because we already know a lot, and especially because we already have an array of basic abstract concepts, thanks to evolution. Like scientists, we can use those concepts to formulate hypotheses about the world. Then, instead of trying to extract patterns from the raw data, we can make predictions about what the data should look like if those hypotheses are right. Along with Plato, such "rationalist" philosophers and psychologists as Descartes and Noam Chomsky took this approach.

Here's an everyday example that illustrates the difference between the two methods: solving the spam plague. The data consist of a long, unsorted list of messages in your in-box. The reality is that some of these messages are genuine and some are spam. How can you use the data to discriminate between them?

Consider the bottom-up technique first. You notice that the spam messages tend to have particular features: a long list of addressees, origins in Nigeria, references to million-dollar prizes, or Viagra. The trouble is that perfectly useful messages might have these features, too. If you looked at enough examples of spam and nonspam emails, you might see not only that spam emails tend to have those features but that the features tend to go together in particular ways (Nigeria plus a million dollars spells trouble). In fact, there might be some subtle higher-level correlations that discriminate the spam messages from the useful ones— a particular pattern of misspellings and IP addresses, say. If you detect those patterns, you can filter out the spam.

The bottom-up machine-learning techniques do just this. The learner gets millions of examples, each with some set of features and each labeled as spam (or some other category) or not. The computer can extract the pattern of features that distinguishes the two, even if it's quite subtle.

How about the top-down approach? I get an email from the editor of the *Journal of Clinical Biology*. It refers to one of my papers and says that they would like to publish an article by me. No Nigeria, no Viagra, no million dollars; the email doesn't have any of the features of spam. But by using what I already know, and thinking in an abstract way about the process that produces spam, I can figure out that this email is suspicious:

1. I know that spammers try to extract money from people by appealing to human greed.

2. I also know that legitimate "open access" journals have started
   covering their costs by charging authors instead of subscribers,
   and that I don't practice anything like clinical biology.

Put all that together and I can produce a good new hypothesis about
where that email came from. It's designed to sucker academics into paying to "publish" an article in a fake journal. The email was a result of the
same dubious process as the other spam emails, even though it looked
nothing like them. I can draw this conclusion from just one example,
and I can go on to test my hypothesis further, beyond anything in the
email itself, by googling the "editor."

In computer terms, I started out with a "generative model" that includes abstract concepts like greed and deception and describes the process that produces email scams. That lets me recognize the classic
Nigerian email spam, but it also lets me imagine many different kinds of
possible spam. When I get the journal email, I can work backward:
"This seems like just the kind of mail that would come out of a spam-
generating process."

The new excitement about AI comes because AI researchers have
recently produced powerful and effective versions of both of these learning methods. But there is nothing profoundly new about the methods
themselves.

BOTTOM-UP DEEP LEARNING

In the 1980s, computer scientists devised an ingenious way to get computers to detect patterns in data: connectionist, or neural-network, architecture (the "neural" part was, and still is, metaphorical). The
approach fell into the doldrums in the 1990s but has recently been
revived with powerful "deep-learning" methods like Google's Deep-
Mind.

For example, you can give a deep-learning program a bunch of Internet
images labeled "cat," others labeled "house," and so on. The program

can detect the patterns differentiating the two sets of images and use that information to label new images correctly. Some kinds of machine learning, called unsupervised learning, can detect patterns in data with no labels at all; they simply look for clusters of features—what scientists call a factor analysis. In the deep-learning machines, these processes are repeated at different levels. Some programs can even discover relevant features from the raw data of pixels or sounds; the computer might begin by detecting the patterns in the raw image that correspond to edges and lines and then find the patterns in those patterns that correspond to faces, and so on.

Another bottom-up technique with a long history is reinforcement learning. In the 1950s, B. F. Skinner, building on the work of John Watson, famously programmed pigeons to perform elaborate actions—even guiding air-launched missiles to their targets (a disturbing echo of recent AI) by giving them a particular schedule of rewards and punishments. The essential idea was that actions that were rewarded would be repeated and those that were punished would not, until the desired behavior was achieved. Even in Skinner's day, this simple process, repeated over and over, could lead to complex behavior. Computers are designed to perform simple operations over and over on a scale that dwarfs human imagination, and computational systems can learn remarkably complex skills in this way.

For example, researchers at Google's DeepMind used a combination of deep learning and reinforcement learning to teach a computer to play Atari video games. The computer knew nothing about how the games worked. It began by acting randomly and got information only about what the screen looked like at each moment and how well it had scored. Deep learning helped interpret the features on the screen, and reinforcement learning rewarded the system for higher scores. The computer got very good at playing several of the games, but it also completely bombed on others that were just as easy for humans to master.

A similar combination of deep learning and reinforcement learning has enabled the success of DeepMind's AlphaZero, a program that managed to beat human players at both chess and Go, equipped with only a basic knowledge of the rules of the game and some planning capacities. AlphaZero has another interesting feature: It works by playing hundreds of millions of games against itself. As it does so, it prunes mistakes that led to losses, and it repeats and elaborates on strategies that led to wins. Such systems, and others involving techniques called generative adversarial networks, generate data as well as observing data.

When you have the computational power to apply those techniques to very large data sets or millions of email messages, Instagram images, or voice recordings, you can solve problems that seemed very difficult before. That's the source of much of the excitement in computer science. But it's worth remembering that those problems—like recognizing that an image is a cat or a spoken word is Siri—are trivial for a human toddler. One of the most interesting discoveries of computer science is that problems that are easy for us (like identifying cats) are hard for computers—much harder than playing chess or Go. Computers need millions of examples to categorize objects that we can categorize with just a few. These bottom-up systems can generalize to new examples; they can label a new image as a cat fairly accurately over all. But they do so in ways quite different from how humans generalize. Some images almost identical to a cat image won't be identified by us as cats at all. Others that look like a random blur will be.

TOP-DOWN BAYESIAN MODELS

The top-down approach played a big role in early AI, and in the 2000s it, too, experienced a revival, in the form of probabilistic, or Bayesian, generative models.

The early attempts to use this approach faced two kinds of problems. First, most patterns of evidence might in principle be explained by many different hypotheses: It's possible that my journal email message is genuine, it just doesn't seem likely. Second, where do the concepts that the generative models use come from in the first place? Plato and Chomsky said you were born with them. But how can we explain how we learn the latest concepts of science? Or how even young children understand about dinosaurs and rocket ships?

Bayesian models combine generative models and hypothesis testing with probability theory, and they address these two problems. A Bayesian model lets you calculate just how likely it is that a particular hypothesis is true, given the data. And by making small but systematic tweaks to the models we already have, and testing them against the data, we can sometimes make new concepts and models from old ones. But these advantages are offset by other problems. The Bayesian techniques can help you choose which of two hypotheses is more likely, but there are almost always an enormous number of possible hypotheses, and no system can efficiently consider them all. How do you decide which hypotheses are worth testing in the first place?

Brenden Lake at NYU and colleagues have used these kinds of top-down methods to solve another problem that's easy for people but extremely difficult for computers: recognizing unfamiliar handwritten characters. Look at a character on a Japanese scroll. Even if you've never seen it before, you can probably tell if it's similar to or different from a character on another Japanese scroll. You can probably draw it and even design a fake Japanese character based on the one you see—one that will look quite different from a Korean or Russian character.*

The bottom-up method for recognizing handwritten characters is to give the computer thousands of examples of each one and let it pull out the salient features. Instead, Lake et al. gave the program a general model of how you draw a character: A stroke goes either right or left; after you finish one, you start another; and so on. When the program saw a particular character, it could infer the sequence of strokes that were most likely to have led to it—just as I inferred that the spam process led to my dubious email. Then it could judge whether a new character was likely to result from that sequence or from a different one, and it could produce a similar set of strokes itself. The program worked much

---

* Brenden M. Lake, Ruslan Salakhutdinov, and Joshua B. Tenenbaum, "Human-Level Concept Learning Through Probabilistic Program Induction," *Science* 350, no. 6266 (2015): 1332–38.

better than a deep-learning program applied to exactly the same data, and it closely mirrored the performance of human beings.

These two approaches to machine learning have complementary strengths and weaknesses. In the bottom-up approach, the program doesn't need much knowledge to begin with, but it needs a great deal of data, and it can generalize only in a limited way. In the top-down approach, the program can learn from just a few examples and make much broader and more varied generalizations, but you need to build much more into it to begin with. A number of investigators are currently trying to combine the two approaches, using deep learning to implement Bayesian inference.

The recent success of AI is partly the result of extensions of those old ideas. But it has more to do with the fact that, thanks to the Internet, we have much more data, and thanks to Moore's Law we have much more computational power to apply to that data. Moreover, an unappreciated fact is that the data we do have has already been sorted and processed by human beings. The cat pictures posted to the Web are canonical cat pictures—pictures that humans have already chosen as "good" pictures. Google Translate works because it takes advantage of millions of human translations and generalizes them to a new piece of text, rather than genuinely understanding the sentences themselves.

But the truly remarkable thing about human children is that they somehow combine the best features of each approach and then go way beyond them. Over the past fifteen years, developmentalists have been exploring the way children learn structure from data. Four-year-olds can learn by taking just one or two examples of data, as a top-down system does, and generalizing to very different concepts. But they can also learn new concepts and models from the data itself, as a bottom-up system does.

For example, in our lab we give young children a "blicket detector"—a new machine to figure out, one they've never seen before. It's a box that lights up and plays music when you put certain objects on it but not others. We give children just one or two examples of how the machine works, showing them that, say, two red blocks make it go, while a green-and-yellow combination doesn't. Even eighteen-month-olds immediately figure out the general principle that the two objects have to be the same to make it go, and they generalize that principle to new examples: For instance, they will choose two objects that have the same *shape* to make the machine work. In other experiments, we've shown that children can even figure out that some hidden invisible property makes the machine go, or that the machine works on some abstract logical principle.*

You can show this in children's everyday learning, too. Young children rapidly learn abstract intuitive theories of biology, physics, and psychology in much the way adult scientists do, even with relatively little data.

The remarkable machine-learning accomplishments of the recent AI systems, both bottom-up and top-down, take place in a narrow and well-defined space of hypotheses and concepts—a precise set of game pieces and moves, a predetermined set of images. In contrast, children and scientists alike sometimes change their concepts in radical ways, performing paradigm shifts rather than simply tweaking the concepts they already have.

Four-year-olds can immediately recognize cats and understand words, but they can also make creative and surprising new inferences that go far beyond their experience. My own grandson recently explained, for example, that if an adult wants to become a child again, he should try not eating any healthy vegetables, since healthy vegetables make a child grow into an adult. This kind of hypothesis, a plausible one that no grown-up would ever entertain, is characteristic of young children. In fact, my colleagues and I have shown systematically that preschoolers are better at coming up with unlikely hypotheses than older children and adults.† We have almost no idea how this kind of creative learning and innovation is possible.

Looking at what children do, though, may give programmers useful

---

* A. Gopnik, T. Griffiths, and C. Lucas, "When Younger Learners Can Be Better (or at Least More Open-Minded) Than Older Ones," *Current Directions in Psychological Science* 24, no. 2 (2015): 87–92.
† A. Gopnik et al., "Changes in Cognitive Flexibility and Hypothesis Search Across Human Life History from Childhood to Adolescence to Adulthood," *PNAS* 114, no. 30 (2017): 7892–99.

hints about directions for computer learning. Two features of children's learning are especially striking. Children are active learners; they don't just passively soak up data like AIs do. Just as scientists experiment, children are intrinsically motivated to extract information from the world around them through their endless play and exploration. Recent studies show that this exploration is more systematic than it looks and is well adapted to find persuasive evidence to support hypothesis formation and theory choice.* Building curiosity into machines and allowing them to actively interact with the world might be a route to more realistic and wide-ranging learning.

Second, children, unlike existing AIs, are social and cultural learners. Humans don't learn in isolation but avail themselves of the accumulated wisdom of past generations. Recent studies show that even preschoolers learn through imitation and by listening to the testimony of others. But they don't simply passively obey their teachers. Instead they take in information from others in a remarkably subtle and sensitive way, making complex inferences about where the information comes from and how trustworthy it is and systematically integrating their own experiences with what they are hearing.†

"Artificial intelligence" and "machine learning" sound scary. And in some ways they are. These systems are being used to control weapons, for example, and we really should be scared about that. Still, natural stupidity can wreak far more havoc than artificial intelligence; we humans will need to be much smarter than we have been in the past to properly regulate the new technologies. But there is not much basis for either the apocalyptic or the utopian vision of AIs replacing humans. Until we solve the basic paradox of learning, the best artificial intelligences will be unable to compete with the average human four-year-old.

---

* L. Schulz, "The Origins of Inquiry: Inductive Inference and Exploration in Early Childhood," *Trends in Cognitive Sciences* 16, no. 7 (2012): 382–89.
† A. Gopnik, *The Gardener and the Carpenter* (New York: Farrar, Straus & Giroux, 2016), chapters 4 and 5.

Chapter 22

ALGORISTS DREAM OF OBJECTIVITY

PETER GALISON

Peter Galison is a science historian; Joseph Pellegrino University Professor and co-founder of the Black Hole Initiative at Harvard University; and the author of Einstein's Clocks, Poincaré's Maps: Empires of Time.

Peter Galison's focus as a science historian is—speaking roughly—on the intersection of theory with experiment.

"For quite a number of years I have been guided in my work by the odd confrontation of abstract ideas and extremely concrete objects," he once told me, in explaining how he thinks about what he does. At the Washington, Connecticut, meeting he discussed the Cold War tension between engineers (like Wiener) and the administrators of the Manhattan Project (like Oppenheimer): "When [Wiener] warns about the dangers of cybernetics, in part he's trying to compete against the kind of portentous language that people like Oppenheimer [used]: 'When I saw the explosion at Trinity, I thought of the Bhagavad Gita—I am death, destroyer of worlds.' That sense, that physics could stand and speak to the nature of the universe and air force policy, was repellent and seductive. In a way, you can see that over and over again in the last decades—nanosciences, recombinant DNA, cybernetics: 'I stand reporting to you on the science that has the promise of salvation and the danger of annihilation—and you should pay attention, because this could kill you.' It's a very seductive narrative, and it's repeated in artificial intelligence and robotics."

As a twenty-four-year old, when I first encountered Wiener's ideas and met his colleagues at the MIT meeting I describe in the book's introduction, I was hardly interested in Wiener's warnings or admonitions. What drove my curiosity was the stark, radical nature of his view of life, based on the mathematical theory of communications in which the message was nonlinear: According to Wiener, "new concepts of communication and control involved a new interpretation of man, of man's knowledge of the universe, and of society." And that led to my first book, which took information theory—the mathematical theory of communications—as a model for all human experience.

In a recent conversation, Peter told me he was beginning to write a book—about building, crashing, and thinking—that considers the black-box nature of cybernetics and how it represents what he thinks of as "the fundamental transformation of learning, machine learning, cybernetics, and the self."

I n his second-best book, the great medieval mathematician al-Khwarizmi described the new place-based Indian form of arithmetic. His name, soon sonically linked to "algorismus" (in late medieval Latin), came to designate procedures acting upon numbers—eventually wending its way through "algorithm" (on the model of "logarithm"), into French, and on into English. But I like the idea of a modern algorist, even if my spellcheck does not. I mean by it someone profoundly suspicious of the intervention of human judgment, someone who takes that judgment to violate the fundamental norms of what it is to be objective (and therefore scientific).

Near the end of the 20th century, a paper by two University of Minnesota psychologists summarized a vast literature that had long roiled the waters of prediction. One side, they judged, had for all too long held resolutely—and ultimately unethically—to the "clinical method" of prediction, which prized all that was *subjective*: "informal," "in the head," and "impressionistic." These clinicians were people (so said the psychologists) who thought they could study their subjects with meticulous care, gather in committees, and make judgment-based predictions about criminal recidivism, college success, medical outcomes, and the like. The other side, the psychologists continued, embodied everything the clinicians did not, embracing the *objective*: "formal," "mechanical," "algorithmic." This the authors took to stand at the root of the whole triumph of post-Galilean science. Not only did science benefit from the actuarial; to a great extent, science *was* the mechanical-actuarial. Breezing through 136 studies of predictions, across domains from sentencing to psychiatry, the authors showed that in 128 of them, predictions using actuarial tables, a multiple-regression equation, or an algorithmic judgment equaled or exceeded in accuracy those using the subjective approach.

They went on to catalog seventeen fallacious justifications for clinging to the clinical. There were the self-interested foot draggers who feared losing their jobs to machines. Others lacked the education to follow statistical arguments. One group mistrusted the formalization of mathematics; another excoriated what they took to be the actuarial "dehumanizing"; yet others said that the aim was to understand, not to predict. But whatever the motivations, the review concluded that it was downright immoral to withhold the power of the objective over the subjective, the algorithmic over expert judgment.*

The algorist view has gained strength. Anne Milgram served as attorney general of the state of New Jersey from 2007 to 2010. When she took office, she wanted to know who the state was arresting, charging, and jailing, and for what crimes. At the time, she reports in a later TED Talk, she could find almost no data or analytics. By imposing statistical prediction, she continues, law enforcement in Camden during her tenure was able to reduce murders by 41 percent, saving thirty-seven lives, while dropping the total crime rate by 26 percent. After joining the Arnold Foundation as its vice president for criminal justice, she established a team of data scientists and statisticians to create a risk-assessment tool; fundamentally, she construed the team's mission as deciding how to put "dangerous people" in jail while releasing the nondangerous. "The reason for this," Milgram contended, "is the way we make decisions. Judges have the best intentions when they make these decisions about risk, but they're making them subjectively. They're like the baseball scouts twenty

---

* William M. Grove and Paul E. Meehl, "Comparative Efficiency of Informal (Subjective, Impressionistic) and Formal (Mechanical, Algorithmic) Prediction Procedures: The Clinical-Statistical Controversy," *Psychology, Public Policy, and Law* 2, no. 2 (1996): 293–323.

years ago who were using their instinct and their experience to try to decide what risk someone poses. They're being subjective, and we know what happens with subjective decision making, which is that we are often wrong." Her team established nine-hundred-plus risk factors, of which nine were most predictive. The most urgent questions for the team were: Will a person commit a new crime? Will that person commit a violent act? Will someone come back to court? We need, concluded Milgram, an "objective measure of risk" that should be inflected by judges' judgment. We know the algorithmic statistical process works. That, she says, is "why Google is Google" and why moneyball wins games.*

Algorists have triumphed. We have grown accustomed to the idea that protocols and data can and should guide us in everyday action, from reminders about where we probably want to go next to the likely occurrence of crime. By now, according to the literature, the legal, ethical, formal, and economic dimensions of algorithms are all quasi-infinite. I'd like to focus on one particular siren song of the algorithm: its promise of objectivity.

Scientific objectivity has a history. That might seem surprising. Isn't the notion—expressed above by the Minnesota psychologists—right? Isn't objectivity co-extensive with science itself? Here it's worth stepping back to reflect on all the epistemic virtues we might value in scientific work. Quantification seems like a good thing to have; so, too, do prediction, explanation, unification, precision, accuracy, certainty, and pedagogical utility. In the best of all possible worlds these epistemic virtues would all pull in the same direction. But they do not—not any more than our ethical virtues necessarily coincide. Rewarding people according to their need may very well conflict with rewarding people according to their ability. Equality, fairness, meritocracy—ethics, in a sense, is all about the adjudication of conflicting goods. Too often we forget that this conflict exists in science, too. Design an instrument to be as sensitive as possible and it often fluctuates wildly, making repetition of a measurement impossible.

---

* TED Talk, January 2014, https://www.ted.com/speakers/anne_milgram.

"Scientific objectivity" entered both the practice and the nomencla-
ture of science after the first third of the 19th century. One sees this
clearly in the scientific atlases that provided scientists with the basic objects of their specialty: There were (and are) atlases of the hand, atlases
of the skull, atlases of clouds, crystals, flowers, bubble-chamber pictures, nuclear emulsions, and diseases of the eye. In the 18th century, it
was obvious that you would not depict this particular, sun-scorched,
caterpillar-chewed clover found outside your house in an atlas. No, you
aimed—if you were a genius natural philosopher like Goethe, Albinus,
or Cheselden—to observe nature but then to perfect the object in question, to abstract it visually to the ideal. Take a skeleton, view it through
a camera lucida, draw it with care. Then correct the "imperfections."
The advantage of this parting of the curtains of mere experience was
clear: It provided a universal guide, one not attached to the vagaries of
individual variation.

As the sciences grew in scope, and scientists grew in number, the
downside of idealization became clearer. It was one thing to have Goethe
depict the "ur-plant" or "ur-insect." It was quite another to have myriad
different scientists fixing their images in different and sometimes contradictory ways. Gradually, from around the 1830s forward, one begins
to see something new: a claim that the image making was done with a
minimum of human intervention, that protocols were followed. This
could mean tracing a leaf with a pencil or pressing it into ink that was
transferred to the page. It meant, too, that one suddenly was proud of
depicting the view through a microscope of a natural object even with its
imperfections. This was a radical idea: snowflakes shown without perfect hexagonal symmetry, color distortion near the edge of a microscope
lens, tissue torn around the edges in the process of its preparation.

Scientific objectivity came to mean that our representations of things
were executed by holding back from intervention—even if it meant reproducing the yellow color near the edge of the image under the microscope despite the fact that the scientist *knew* that the discoloration was
from the lens, not a feature of the object of inquiry. The advantage of
objectivity was clear: It superseded the desire to see a theory realized or

a generally accepted view confirmed. But objectivity came at a cost. You lost that precise, easily teachable, colored, full depth-of-field artist's rendition of a dissected corpse. You got a blurry, bad depth-of-field, black-and-white photograph that no medical student (nor even many medical colleagues) could use to learn and compare cases. Still, for a long stretch of the 19th century, the virtue of hands-off, self-restraining objectivity was on the rise.

Starting in the 1930s, the hardline scientific objectivity in scientific representation began running into trouble. In cataloging stellar spectra, for example, no algorithm could compete with highly trained observers who could sort them with far greater accuracy and replicability than any purely rule-following procedure. By the late 1940s, doctors had begun learning how to read electroencephalograms. Expert judgment was needed to sort out different kinds of seizure readings, while none of the early attempts to use frequency analysis could match that judgment. Solar magnetograms—mapping the magnetic fields across the sun—required the trained expert to pry the real signal from artifacts that emerged from the measuring instruments. Even particle physicists recognized that they could not program a computer to sort certain kinds of tracks into the right bins; trained judgment was needed.

There should be no confusion here: This was not a return to the invoked genius of an 18th-century idealizer. No one thought you could train to be a Goethe who alone among scientists could pick out the universal, ideal form of a plant, insect, or cloud. Expertise could be learned—you could take a course to learn to make expert judgments about electroencephalograms, stellar spectra, or bubble-chamber tracks; alas, no one has ever thought you could take a course that would lead to the mastery of exceptional insight. There can be no royal road to becoming Goethe. In scientific atlas after scientific atlas, one sees explicit argument that "subjective" factors had to be part of the scientific work needed to create, classify, and interpret scientific images.

What we see in so many of the algorists' claims is a tremendous desire to find scientific objectivity precisely by abandoning judgment and relying on mechanical procedures—in the name of scientific objectivity.

Many American states have legislated the use of sentencing and parole algorithms. Better a machine, it is argued, than the vagaries of a judge's judgment.

So here is a warning from the sciences. Hands-off algorithmic proceduralism did indeed have its heyday in the 19th century, and of course still plays a role in many of the most successful technical and scientific endeavors. But the idea that mechanical objectivity, construed as binding self-restraint, follows a simple, monotonic curve increasing from the bad impressionistic clinician to the good externalized actuary simply does not answer to the more interesting and nuanced history of the sciences.

There is a more important lesson from the sciences. Mechanical objectivity is a scientific virtue among others, and the hard sciences learned that lesson often. We must do the same in the legal and social scientific domains. What happens, for example, when the secret, proprietary algorithm sends one person to prison for ten years and another for five years for the same crime? Rebecca Wexler, visiting fellow at the Yale Law School Information Society Project, has explored that question, and the tremendous cost that trade-secret algorithms impose on the possibility of a fair legal defense.* Indeed, for a variety of reasons, law enforcement may not want to share the algorithms used to make DNA, chemical, or fingerprint identifications, which puts the defense in a much-weakened position to make its case. In the courtroom, objectivity, trade secrets, and judicial transparency may pull in opposite directions. It reminds me of a moment in the history of physics. Just after World War II, the film giants Kodak and Ilford perfected a film that could be used to reveal the interactions and decays of elementary particles. The physicists were thrilled, of course—until the film companies told them that the composition of the film was a trade secret, so the scientists would never gain complete confidence that they understood the processes they were studying. Proving things with unopenable black boxes can be a dangerous game for scientists, and doubly so for criminal justice.

---

* Rebecca Wexler, "Life, Liberty, and Trade Secrets: Intellectual Property in the Criminal Justice System," *Stanford Law Review* 70 (2018).

Other critics have underscored how perilous it is to rely on an accused (or convicted) person's address or other variables that can easily become, inside the black box of algorithmic sentencing, a proxy for race. By dint of everyday experience, we have grown used to the fact that airport security is different for children under the age of twelve and adults over the age of seventy-five. What factors do we want the algorists to have in their often hidden procedures? Education? Income? Employment history? What one has read, watched, visited, or bought? Prior contact with law enforcement? How do we want algorists to weight those factors? Predictive analytics predicated on mechanical objectivity comes at a price. Sometimes it may be a price worth paying; sometimes that price would be devastating for the just society we want to have.

More generally, as the convergence of algorithms and Big Data governs a greater and greater part of our lives, it would be well worth keeping in mind these two lessons from the history of the sciences: Judgment is not the discarded husk of a now pure objectivity of self-restraint. And mechanical objectivity is a virtue competing among others, not the defining essence of the scientific enterprise. They are lessons to bear in mind, even if algorists dream of objectivity.

Chapter 23

THE RIGHTS OF MACHINES

GEORGE M. CHURCH

George M. Church *is Robert Winthrop Professor of Genetics at Harvard Medical School, Professor of Health Sciences and Technology at Harvard-MIT, and co-author (with Ed Regis) of* Regenesis: How Synthetic Biology Will Reinvent Nature and Ourselves.

In the past decade, genetic engineering has caught up with computer science with regard to how new scientific initiatives are shaping our lives. Genetic engineer **George Church**, a pioneer of the revolution in reading and writing biology, is central to this new landscape of ideas. He thinks of the body as an operating system, with engineers taking the place of traditional biologists in retooling stripped-down components of organisms (from atoms to organs) in much the same vein as in the late 1970s, when electrical engineers were working their way to the first personal computer by assembling circuit boards, hard drives, monitors, etc. George created and is director of the Personal Genome Project, which provides the world's only open-access information on human genomic, environmental, and trait data (GET) and sparked the growing DNA ancestry industry.

He was instrumental in laying the groundwork for President Obama's 2013 BRAIN (Brain Research through Advancing Innovative Neurotechnologies) Initiative—in aid of improving the brains of human beings to the point where, for much of what sustains us, we might not need the help of (potentially dicey) AIs. "It could be that some of the BRAIN Initiative projects allow us to build human brains that are more consistent with our ethics and capable of doing advanced tasks like artificial intelligence," George has said. "The safest path by far is getting humans to do all the tasks that they would like to delegate to machines, but we're not yet firmly on that super-safe path."

More recently, his crucially important pioneering use of the enzyme CRISPR (as well as methods better than CRISPR) to edit the genes of human cells is sometimes missed by the media in the telling of the CRISPR origins story.

George's attitude toward future forms of artificial general intelligence is friendly, as evinced in the essay that follows. At the same time, he never loses sight of the AI-safety issue. On that subject, he recently remarked: "The main risk in AI, to my mind, is not so much whether we can mathematically understand what they're thinking; it's whether we're capable of teaching them ethical behavior. We're barely capable of teaching each other ethical behavior."

n 1950, Norbert Wiener's *The Human Use of Human Beings* was at the cutting edge of vision and speculation in proclaiming that

> the machine like the djinnee, which can learn and can make decisions on the basis of its learning, will in no way be obliged to make such decisions as we should have made, or will be acceptable to us. . . . Whether we entrust our decisions to machines of metal, or to those machines of flesh and blood which are bureaus and vast laboratories and armies and corporations . . . [t]he hour is very late, and the choice of good and evil knocks at our door.

But this was his book's denouement, and it has left us hanging now for sixty-eight years, lacking not only prescriptions and proscriptions but even a well-articulated "problem statement." We have since seen similar warnings about the threat of our machines, even in the form of outreach to the masses, via films like *Colossus: The Forbin Project* (1970), *The Terminator* (1984), *The Matrix* (1999), and *Ex Machina* (2015). But now the time is ripe for a major update, with fresh, new perspectives—notably focused on generalizations of our "human" rights and our existential needs.

Concern has tended to focus on "us versus them [robots]" or "grey goo [nanotech]" or "monocultures of clones [bio]." To extrapolate current trends: What if we could make or grow almost anything and engineer any level of safety and efficacy desired? Any thinking being (made of any arrangement of atoms) could have access to any technology.

Probably we should be less concerned about us-versus-them and more concerned about the rights of all sentients in the face of an emerging unprecedented diversity of minds. We should be harnessing this diversity to minimize global existential risks, like supervolcanoes and asteroids.

But should we say "should"? (Disclaimer: In this and many other cases, when a technologist describes a societal path that "could," "would," or "should" happen, this doesn't necessarily equate to the preferences of the author. It could reflect warning, uncertainty, and/or detached assessment.) Roboticist Gianmarco Veruggio and others have raised issues of roboethics since 2002; the U.K. Department of Trade and Industry and the RAND spin-off Institute for the Future have raised issues of robot rights since 2006.

"IS VERSUS OUGHT"

It is commonplace to say that science concerns "is," not "ought." Stephen Jay Gould's "non-overlapping magisteria" view argues that facts must be completely distinct from values. Similarly, the 1999 document *Science and Creationism* from the U.S. National Academy of Sciences noted that "science and religion occupy two separate realms." This division has been critiqued by evolutionary biologist Richard Dawkins, me, and others. We can discuss "should" if framed as "we should do X in order to achieve Y." Which Y should be a high priority is not necessarily settled by democratic vote but might be settled by Darwinian vote. Value systems and religions wax and wane, diversify, diverge, and merge just as living species do: subject to selection. The ultimate "value" (the "should") is survival of genes and memes.

Few religions say that there is no connection between our physical being and the spiritual world. Miracles are documented. Conflicts between Church doctrine and Galileo and Darwin are eventually resolved. Faith and ethics are widespread in our species and can be studied using scientific methods, including but not limited to fMRI, psychoactive drugs, questionnaires, etc.

Very practically, we have to address the ethical rules that should be built in, learned, or probabilistically chosen for increasingly intelligent and diverse machines. We have a whole series of Trolley Problems. At what number of people in line for death should the computer decide to shift a moving trolley to one person? Ultimately this might be a deep-learning problem—one in which huge databases of facts and contingencies can be taken into account, some seemingly far from the ethics at hand.

For example, the computer might infer that the person who would escape death if the trolley is left alone is a convicted terrorist recidivist loaded up with doomsday pathogens, or a saintly POTUS—or part of a much more elaborate chain of events in detailed alternative realities. If one of these problem descriptions seems paradoxical or illogical, it may be that the authors of the Trolley Problem have adjusted the weights on each side of the balance such that hesitant indecision is inevitable.

Alternatively, one can use misdirection to rig the system, such that the error modes are not at the level of attention. For example, in the Trolley Problem, the real ethical decision was made years earlier when pedestrians were given access to the rails—or even before that, when we voted to spend more on entertainment than on public safety. Questions that at first seem alien and troubling, like "Who owns the new minds, and who pays for their mistakes?" are similar to well-established laws about who owns and pays for the sins of a corporation.

THE SLIPPERY SLOPES

We can (over)simplify ethics by claiming that certain scenarios won't happen. The technical challenges or the bright red lines that cannot be crossed are reassuring, but the reality is that once the benefits seem to outweigh the risks (even briefly and barely), the red lines shift. Just before Louise Brown's birth in 1978, many people were worried that she "would turn out to be a little monster, in some way, shape or form, deformed, something wrong with her."* Few would hold this view of in-vitro fertilization today.

What technologies are lubricating the slope toward multiplex sentience? It is not merely deep machine-learning algorithms with Big Iron. We have engineered rodents to be significantly better at a variety of cognitive tasks as well as to exhibit other relevant traits, such as persistence and low anxiety. Will this be applicable to animals that are already at the door of humanlike intelligence? Several show self-recognition in a mirror test—chimpanzees, bonobos, orangutans, some dolphins and whales, and magpies.

Even the bright red line for human manipulation of human beings shows many signs of moving or breaking completely. More than twenty-three hundred approved clinical trials for gene therapy are in progress worldwide. A major medical goal is the treatment or prevention of cognitive decline, especially in light of our rapidly aging global demographic. Some treatments of cognitive decline will include cognitive enhancements (drugs, genes, cells, transplants, implants, and so on). These will be used off-label. The rules of athletic competition (e.g., banning augmentation with steroids or erythropoietin) do not apply to intellectual competition in the real world. Every bit of progress on cognitive decline is in play for off-label use.

Another frontier of the human use of humans is "brain organoids." We can now accelerate developmental biology. Processes that normally take months can happen in four days in the lab using the right recipes of transcription factors. We can make brains that, with increasing fidelity, recapitulate the differences between people born with aberrant cognitive abilities (e.g., microcephaly). Proper vasculature (veins, arteries, and capillaries) missing from earlier successes are now added, enabling brain

---

* "Then, Doctors 'All Anxious' About Test-tube Baby," http://edition.cnn.com/2003/HEALTH/parenting/07/25/cnna.copperman.

organoids to surpass the former submicroliter limit to possibly exceed the 1.2-liter size of modern human brains (or even the 5-liter elephant or 8-liter sperm whale brains).

CONVENTIONAL COMPUTERS VERSUS BIO-ELECTRONIC HYBRIDS

As Moore's Law miniaturization approaches its next speed bump (surely not a solid wall), we see the limits of the stochastics of dopant atoms in silicon slabs and the limits of beam-fabrication methods at around 10-nanometer feature size. Power (energy consumption) issues are also apparent: The great Watson, winner of *Jeopardy!*, used 85,000 watts real time, while the human brains were using 20 watts each. To be fair, the human body needs 100 watts to operate and twenty years to build, hence about 6 trillion joules of energy to "manufacture" a mature human brain. The cost of manufacturing Watson-scale computing is similar. So why aren't humans displacing computers?

For one, the *Jeopardy!* contestants' brains were doing far more than information retrieval—much of which would be considered mere distractions by Watson (e.g., cerebellar control of smiling). Other parts allow leaping out of the box with transcendence unfathomable by Watson, such as what we see in Einstein's five *annus mirabilis* papers of 1905. Also, humans consume more energy than the minimum (100 watts) required for life and reproduction. People in India use an average of 700 watts per person; it's 10,000 watts in the U.S. Both are still less than the 85,000 watts Watson uses. Computers can become more like us via neuromorphic computing, possibly a thousandfold. But human brains could get more efficient, too. The organoid brain-in-a-bottle could get closer to the 20 watts limit. The idiosyncratic advantages of computers for math, storage, and search, faculties of limited use to our ancestors, could be designed and evolved anew in labs.

Facebook, the National Security Agency, and others are constructing exabyte-scale storage facilities at more than a megawatt and four hectares, while DNA can store that amount in a milligram. Clearly, DNA is not a mature storage technology, but with Microsoft and Technicolor doubling down on it, we would be wise to pay attention. The main reason for the 6 trillion joules of energy required to get a productive human mind is the twenty years required for training.

Even though a supercomputer can "train" a clone of zemself in seconds, the energy cost of producing a mature silicon clone is comparable. Engineering (*Homo*) prodigies might make a small impact on this slow process, but speeding up development and implanting extensive memory (as DNA-exabytes or other means) could reduce duplication time of a bio-computer to close to the doubling time of cells (ranging from eleven minutes to twenty-four hours). The point is that while we may not know what ratio of bio/homo/nano/robo hybrids will be dominant at each step of our accelerating evolution, we can aim for high levels of humane, fair, and safe treatment ("use") of one another.

Bills of Rights date back to 1689 in England. FDR proclaimed the "Four Freedoms"—freedom of speech, freedom of conscience, freedom from fear, and freedom from want. The U.N.'s Universal Declaration of Human Rights in 1948 included the right to life; the prohibition of slavery; defense of rights when violated; freedom of movement; freedom of association, thought, conscience, and religion; social, economic, and cultural rights; duties of the individual to society; and prohibition of use of rights in contravention of the purposes and principles of the United Nations.

The "universal" nature of these rights is not universally embraced and is subject to extensive critique and noncompliance. How does the emergence of non-*Homo* intelligences affect this discussion? At a minimum, it is becoming rapidly difficult to hide behind vague intuition for ethical decisions—"I know it when I see it" (U.S. Supreme Court Justice Potter Stewart, 1964) or the "wisdom of repugnance" (aka "yuck factor," Leon Kass, 1997), or vague appeals to "common sense." As we have to deal with minds alien to us, sometimes quite literal from our viewpoint, we need to be explicit—yea, even algorithmic.

Self-driving cars, drones, stock-market transactions, NSA searches, etc., require rapid, preapproved decision making. We may gain insights into many aspects of ethics that we have been trying to pin down and explain for centuries. The challenges have included conflicting priorities, as well as ingrained biological, sociological, and semilogical cognitive biases. Notably far from consensus in universal dogmas about human rights are notions of privacy and dignity, even though these influence many laws and guidelines.

Humans might want the right to march in to read (and change) the minds of computers to see why they're making decisions at odds with our (*Homo*) instincts. Is it not fair for machines to ask the same of us? We note the growth of movements toward transparency in potential financial conflicts; "open-source" software, hardware, and wetware; the Fair Access to Science and Technology Research Act (FASTR); and the Open Humans Foundation.

In his 1976 book *Computer Power and Human Reason*, Joseph Weizenbaum argued that machines should not replace *Homo* in situations requiring respect, dignity, or care, while others (author Pamela McCorduck and computer scientists like John McCarthy and Bill Hibbard) replied that machines can be more impartial, calm, and consistent and less abusive or mischievous than people in such positions.

EQUALITY

What did the thirty-three-year-old Thomas Jefferson mean in 1776 when he wrote "We hold these Truths to be self-evident, that all Men are created equal, that they are endowed by their Creator with certain unalienable Rights, that among these are Life, Liberty, and the Pursuit of Happiness"? The spectrum of current humans is vast. In 1776, "Men" did not include people of color or women. Even today, humans born with congenital cognitive or behavioral issues are destined for unequal (albeit in most cases compassionate) treatment—Down syndrome, Tay-Sachs disease, Fragile X syndrome, cerebral palsy, and so on.

And as we change geographical location and mature, our unequal rights change dramatically. Embryos, infants, children, teens, adults, patients, felons, gender identities and gender preferences, the very rich and very poor—all of these face different rights and socioeconomic realities. One path to new mind-types obtaining and retaining rights similar to the most elite humans would be to keep a *Homo* component, like a human shield or figurehead monarch/CEO, signing blindly enormous technical documents, making snap financial, health, diplomatic, military, or security decisions. We will probably have great difficulty pulling the plug, modifying, or erasing (killing) a computer and its memories—especially if it has befriended humans and made spectacularly compelling pleas for survival (as all excellent researchers fighting for their lives would do).

Even Scott Adams, creator of *Dilbert*, has weighed in on this topic, supported by experiments at Eindhoven University in 2005 noting how susceptible humans are to a robot-as-victim equivalent of the Milgram experiments done at Yale beginning in 1961. Given the many rights of corporations, including ownership of property, it seems likely that other machines will obtain similar rights, and it will be a struggle to maintain inequities of selective rights along multiaxis gradients of intellect and ersatz feelings.

RADICALLY DIVERGENT RULES FOR HUMANS VERSUS NONHUMANS AND HYBRIDS

The divide noted above for intra *Homo sapiens* variation in rights explodes into a riot of inequality as soon as we move to entities that overlap (or will soon) the spectrum of humanity. In Google Street View, people's faces and car license plates are blurred out. Video devices are excluded from many settings, such as courts and committee meetings. Wearable and public cameras with facial-recognition software touch taboos. Should people with hyperthymesia or photographic memories be excluded from those same settings?

Shouldn't people with prosopagnosia (face blindness) or forgetfulness be able to benefit from facial-recognition software and optical character recognition wherever they go, and if them, then why not everyone? If we all have those tools to some extent, shouldn't we all be able to benefit?

These scenarios echo Kurt Vonnegut's 1961 short story "Harrison Bergeron," in which exceptional aptitude is suppressed in deference to the mediocre lowest common denominator of society. Thought experiments like John Searle's Chinese Room and Isaac Asimov's Three Laws of Robotics all appeal to the sorts of intuitions plaguing human brains that Daniel Kahneman, Amos Tversky, and others have demonstrated. The Chinese Room experiment posits that a mind composed of mechanical and *Homo sapiens* parts cannot be conscious, no matter how competent at intelligent human (Chinese) conversation, unless a human can identify the source of the consciousness and "feel" it. Enforced preference for Asimov's First and Second Laws favor human minds over any other mind meekly present in his Third Law, of self-preservation.

If robots don't have exactly the same consciousness as humans, then this is used as an excuse to give them different rights, analogous to arguments that other tribes or races are less than human. Do robots already show free will? Are they already self-conscious? The robots Qbo have passed the "mirror test" for self-recognition and the robots NAO have passed a related test of recognizing their own voice and inferring their internal state of being, mute or not.

For free will, we have algorithms that are neither fully deterministic nor random but aimed at nearly optimal probabilistic decision making. One could argue that this is a practical Darwinian consequence of game theory. For many (not all) games/problems, if we're totally predictable or totally random, then we tend to lose.

What is the appeal of free will anyway? Historically it gave us a way to assign blame in the context of reward and punishment on Earth or in the afterlife. The goals of punishment might include nudging the priorities of the individual to assist the survival of the species. In extreme cases, this could include imprisonment or other restrictions, if Skinnerian positive/negative reinforcement is inadequate to protect society.

Clearly, such tools can apply to free will, seen broadly—to any machine whose behavior we'd like to manage.

We could argue as to whether the robot actually experiences subjective qualia for free will or self-consciousness, but the same applies to evaluating a human. How do we know that a sociopath, a coma patient, a person with Williams syndrome, or a baby has the same free will or self-consciousness as our own? And what does it matter, practically? If humans (of any sort) convincingly claim to experience consciousness, pain, faith, happiness, ambition, and/or utility to society, should we deny them rights because their hypothetical qualia are hypothetically different from ours?

The sharp red lines of prohibition, over which we supposedly will never step, increasingly seem to be short-lived and not sensible. The line between human and machines blurs, both because machines become more humanlike and because humans become more machinelike—not only since we increasingly blindly follow GPS scripts, reflex tweets, and carefully crafted marketing, but also as we digest ever more insights into our brain and genetic programming mechanisms. The NIH BRAIN Initiative is developing innovative technologies and using these to map out the connections and activity of mental circuitry so as to improve electronic and synthetic neurobiological ware.

Various red lines depend on genetic exceptionalism, in which genetics is considered permanently heritable (although it is provably reversible), whereas exempt (and lethal) technologies, like cars, are, for all intents and purposes, irreversible due to social and economic forces. Within genetics, a red line makes us ban or avoid genetically modified foods but embrace genetically modified bacteria making insulin, or genetically modified humans—witness mitochondrial therapies approved in Europe for human adults and embryos.

The line for germline manipulation seems less sensible than the usual, practical line drawn at safety and efficacy. Marriages of two healthy carriers of the same genetic disease have a choice between no child of their own, 25 percent loss of embryos via abortion (spontaneous or induced), 80 percent loss via in-vitro fertilization, or potential 0 percent embryo loss via sperm (germline) engineering. It seems premature to declare this last option unlikely.

For "human subject research," we refer to the 1964 Declaration of Helsinki, keeping in mind the 1932–72 Tuskegee syphilis experiment, possibly the most infamous biomedical research study in U.S. history. In 2015, the Nonhuman Rights Project filed a lawsuit with the New York State Supreme Court on behalf of two chimpanzees kept for research by Stony Brook University. The appellate court decision was that chimps are not to be treated as legal persons since they "do not have duties and responsibilities in society," despite Jane Goodall's and others' claim that they do, and despite arguments that such a decision could be applied to children and the disabled.*

What prevents extension to other animals, organoids, machines, and hybrids? As we (e.g., Hawking, Musk, Tallinn, Wilczek, Tegmark) have promoted bans on "autonomous weapons," we have demonized one type of "dumb" machine, while other machines—for instance, those composed of many *Homo sapiens* voting—can be more lethal and more misguided.

Do transhumans roam the Earth already? Consider the "uncontacted peoples," such as the Sentinelese and Andamanese of India, the Korowai of Indonesia, the Mashco-Piro of Peru, the Pintupi of Australia, the Surma of Ethiopia, the Ruc of Vietnam, the Ayoreo-Totobiegosode of Paraguay, the Himba of Namibia, and dozens of tribes in Papua New Guinea. How would they or our ancestors respond? We could define "transhuman" as people and cultures not comprehensible to humans living in a modern, yet untechnological culture.

Such modern Stone Age people would have great trouble understanding why we celebrate the recent LIGO gravity-wave evidence supporting the hundred-year-old general theory of relativity. They would scratch their heads as to why we have atomic clocks, or GPS satellites so we can find our way home, or why and how we have expanded our vision from a narrow optical band to the full spectrum from radio to gamma.

* https://www.nbcnews.com/news/us-news/lawyer-denying-chimpanzees-rights-could-backfire-disabled-n734566.

We can move faster than any other living species; indeed, we can reach escape velocity from Earth and survive in the very cold vacuum of space.

If those characteristics (and hundreds more) don't constitute transhumanism, then what would? If we feel that the judge of transhumanism should not be fully paleo-culture humans but recent humans, then how would we ever reach transhuman status? We "recent humans" may always be capable of comprehending each new technological increment—never adequately surprised to declare arrival at a (moving) transhuman target. The science-fiction prophet William Gibson said, "The future is already here—it's just not very evenly distributed." While this underestimates the next round of "future," certainly millions of us are transhuman already—with most of us asking for more. The question "What was a human?" has already transmogrified into "What were the many kinds of transhumans? . . . And what were their rights?"

Chapter 24

THE ARTISTIC USE OF CYBERNETIC BEINGS

CAROLINE A. JONES

Caroline A. Jones is a professor of art history in the Department of Architecture at MIT and author of Eyesight Alone: Clement Greenberg's Modernism and the Bureaucratization of the Senses, Machine in the Studio: Constructing the Postwar American Artist, *and* The Global Work of Art.

Caroline A. Jones's interest in modern and contemporary art is enriched by a willingness to delve into the technologies involved in its production, distribution, and reception. "As an art historian, a lot of my questions are about what kind of art we can make, what kind of thought we can make, what kind of ideas we can make that could stretch the human beyond our stubborn, selfish, 'only concerned with our small group' parameters. The philosophers and philosophies I'm drawn to are those that question the Western obsession with individualism. Those are coming from so many different places, and they're reviving so many different kinds of questions and problems that were raised in the 1960s."

She has recently turned her attention to the history of cybernetics. Her MIT course, Automata, Automatism, Systems, Cybernetics, explores the history of the human/machine interface in terms of feedback, exploring the cultural rather than engineering uptake of this idea. She begins with primary readings by Wiener, Shannon, and Turing and then pivots from the scientists and engineers to the work and ideas of artists, feminists, and postmodern theorists. Her goal: to come up with a new central paradigm of evolution that's culture based—"communalism and interspecies symbiosis rather than survival of the fittest."

As a historian, Caroline draws a distinction between what she has termed "left cybernetics" and "right cybernetics": "What do I mean by left cybernetics? In one sense, it's a pun or a joke: the cybernetics that was 'left' behind. On another level, it's a vague political grouping connoting our Left Coast: California, Esalen, the group that Dave Kaiser calls the 'hippie physicists.' It's not an adequate term, but it's a way of recognizing that there was a group beholden to the military-industrial complex, sometimes very unhappily, who gave us the tools to critique it."

Cybernated art is very important, but art
for cybernated life is more important.

—NAM JUNE PAIK, 1966

rtificial intelligence was not what artists first wanted out of cybernetics, once Norbert Wiener's *The Human Use of Human Beings:
Cybernetics and Society* came out in 1950. The range of artists who
identified themselves with cybernetics in the fifties and sixties initially
had little access to "thinking machines." Moreover, craft-minded engineers had already been making turtles, jugglers, and light-seeking robot
babes, not giant brains. Using breadboards, copper wire, simple switches,
and electronic sensors, artists followed cyberneticians in making sculptures and environments that simulated interactive sentience—analog
movements and interfaces that had more to do with instinctive drives
and postwar sexual politics than the automation of knowledge production. Now obscured by an ideology of a free-floating "intelligence" un-
tethered by either hardware or flesh, AI has forgotten the early days of
cybernetics' uptake by artists. Those efforts are worth revisiting; they
modeled relations with what the French philosophers Gilles Deleuze
and Félix Guattari have called the "machinic phylum," having to do
with how humans think and feel in bodies engaged with a physical, material, emotionally stimulating, and signaling world.

Cybernetics now seems to have collapsed into an all-pervasive discourse of AI that was far from preordained. "Cybernetics," as a word,
claimed postwar newness for concepts that were easily four centuries

old: notions of feedback, machine damping, biological homeostasis, logical calculation, and systems thinking that had been around since the
Enlightenment (boosted by the Industrial Revolution). The names in
this lineage include Descartes, Leibniz, Sadi Carnot, Clausius, Maxwell,
and Watt. Wiener's coinage nonetheless had profound cultural effects.*
The ubiquity today of the prefix *cyber-* confirms the desire for a crisp
signifier of the tangled relations between humans and machines. In
Wiener's usage, things "cyber" simply involved "control and communication in the animal and the machine." But after the digital revolution,
"cyber" moved beyond servomechanisms, feedback loops, and switches
to encompass software, algorithms, and cyborgs. The work of cybernetically inclined artists concerns the emergent behaviors of life that elude
AI in its current condition.

As to that original coinage, Wiener had reached back to the ancient
Greek to borrow the word for "steersman" (κυβερνήτης/*kubernétés*), a
masculine figure channeling power and instinct at the helm of a ship,
who read the waves, judged the wind, kept a hand on the tiller, and directed the slaves as they mindlessly (mechanically) churned their oars.
The Greek had already migrated into modern English via Latin, going
from *kuber-* to *guber*—the root of "gubernatorial" and "governor," another term for masculine control, deployed by James Watt to describe
his 19th-century device for modulating a runaway steam engine. Cybernetics thus took ideas that had long analogized people and devices and
generalized them to an applied science by adding that "-ics." Wiener's
three *c*'s (command, control, communication) drew on the mathematics
of probability to formalize systems (whether biological or mechanical)
theorized as a set of inputs of information achieving outputs of actions
in an environment—a muscular, fleshy agenda often minimized in gene-
alogies of AI.

But the etymology does little to capture the excitement felt by partici-
pants, as mathematics joined theoretical biology (Arturo Rosenblueth)

* Wiener later had to admit the earlier coinage of the word in 1834 by André-Marie Ampère, who had intended
it to mean the "science of government," a concept that remained dormant until the 20th century.

and information theory (Claude Shannon, Walter Pitts, Warren McCulloch) to produce a barrage of interdisciplinary research and publications viewed as changing not just the way science was done but the way future humans would engage with the technosphere. As Wiener put it, "We have modified our environment so radically that we must now modify ourselves in order to exist."* The pressing question is: How are we modifying ourselves? Are we going in the right direction or have we lost our way, becoming the tools of our tools? Revisiting the early history of humanist/artists' contribution to cybernetics may help direct us toward a less perilous, more ethical future.

The year 1968 was a high-water mark of the cultural diffusion and artistic uptake of the term. In that year, the Howard Wise gallery opened its show of Wen-Ying Tsai's "Cybernetic Sculpture" in Midtown Manhattan, and Polish émigré Jasia Reichardt opened her exhibition "Cybernetic Serendipity" at London's ICA. (The "Cybernetic" in her title was intended to evoke "made by or with computers," even though most of the artworks on view had no computers, as such, in their responsive circuits.) The two decades between 1948 and 1968 had seen both the fanning out of cybernetic concepts into a broader culture and the spread of computation machines themselves in a slow migration from proprietary military equipment, through the multinational corporation, to the academic lab, where access began to be granted to artists. The availability of cybernetic components—"sensor organs" (electronic eyes, motion sensors, microphones) and "effector organs" (electronic breadboards, switches, hydraulics, pneumatics)—on the home hobbyist front rendered the computer less an "electronic brain" than an adjunct organ in a kit of parts. There was not yet a ruling metaphor of "artificial intelligence." So artists were bricoleurs of electronic bodies, interested in actions rather than in calculation or cognition. There were inklings of "computer" as calculator in the drive toward *Homo rationalis*, but more in aspiration than in achievement.

In light of today's digital convergence in art/science imaging tools,

---

* *The Human Use of Human Beings* (Boston: Houghton Mifflin, 1954), 46.

Reichardt's show was prophetic in its insistence on confusing the boundaries between art and what we might dub "creative applied science."
According to the catalog, "no visitor to the exhibition, unless he reads all
the notes relating to all the works, will know whether he is looking at
something made by an artist, engineer, mathematician, or architect." So
the comically dysfunctional robot by Nam June Paik, *Robot K-456* (1964),
featured on the catalog's cover and described as "a female robot known
for her disturbing and idiosyncratic behavior," would face off against a
balletic *Colloquy of Mobiles* (1968) from second-order cybernetician
Gordon Pask. Pask worked with a London theater designer to craft a
spindly "male" apparatus of hinges and rods, set up to communicate
with bulbous "female" fiberglass entities nearby. Whether anyone could
actually map the quiddities of the program (or glean its reactionary
gender theater) without reading the catalog essay is an open question.
What is significant is Pask's focus on the behaviors of his automata, their
interactivity, their responsiveness within an artificially modulated
environment, and their "reflection" of human behaviors.

The ICA's "Cybernetic Serendipity" introduced an important paradigm: the machinic ecosystem, in which the viewer was a biological part,
tasked with figuring out just what the triggers for interaction might be.
The visitors in those London galleries suddenly became "cybernetic
organisms"—cyborgs—since to experience the art adequately, one
needed to enter a kind of symbiotic colloquy with the servomechanisms.
This turn toward human-machine interactive environments as an aesthetic becomes clearer when we examine a few other artworks from the
period, beginning with one constituting an early instance of emergent
behavior—*Senster,* the interactive sculpture by artist/engineer Edward
Ihnatowicz (1970), celebrated by medical robotics engineer Alex Ziva-
novic, editor of a website devoted to Ihnatowicz's little-known career, as
"one of the first computer controlled interactive robotic works of art."
Here, "the computer" makes its entry (albeit a twelve-bit, limited device). But rather than "intelligence," Ihnatowicz sought to make an avatar of affective behavior. Key to *Senster*'s uncanny success was the
programming with which Ihnatowicz constrained the fifteen-foot-long

hydraulic apparatus (its hinge design and looming appearance inspired by a lobster claw) to convey shyness in responding to humans in its proximity. *Senster*'s sound channels and motion sensors were set to recoil at loud noises and sudden aggressive movements. Only those humans willing to speak softly and modulate their gestures would be rewarded by *Senster*'s quiet, inquisitive approach—an experience that became real for Ihnatowicz himself when he first assembled the program and the machine turned to him solicitously after he'd cleared his throat.

In these artistic uses of cybernetic beings, we sense a growing necessity to train the public to experience itself as embedded in a technologized environment, modifying itself to communicate intuitively with machines. This necessity had already become explicit in Tsai's "Cybernetic Sculpture" show. Those experiencing his immersive installation were expected to experiment with machinic life: What behaviors would trigger the servomechanisms? Likely, the human gallery attendant would have had to explain the protocol: "Clap your hands—that gets the sculptures to respond." As an early critic described it:

> A grove of slender stainless-steel rods rises from a plate. This base vibrates at 30 cycles per second; the rods flex rapidly, in harmonic curves. Set in a dark room, they are lit by strobes. The pulse of the flashing lights varies—they are connected to sound and proximity sensors. The result is that when one approaches a Tsai or makes a noise in its vicinity, the thing responds. The rods appear to move; there is a shimmering, a flashing, an eerie ballet of metal, whose apparent movements range from stillness to jittering and back to a slow, indescribably sensuous undulation.*

Like *Senster,* the apparatus stimulated (and simulated) an affective rather than a rational interaction. Humans felt they were encountering behaviors indicative of responsive life; Tsai's entities were often classed as "vegetal" or "aquatic." Such environmental and kinetic ambitions

---

* Robert Hughes, *Time* magazine (October 2, 1972) review of Tsai exhibition at Denise René gallery.

were widespread in the international art world of the time. Beyond the stable at Howard Wise, there were the émigrés forming the collective GRAV in Paris, the "cybernetic architectures" of Nicolas Schöffer, the light and plastic gyrations of the German Zero Gruppe, and so on—all defining and informing the genre of installation art to come.

The artistic use of cybernetic beings in the late sixties made no investment in "intelligence." Knowing machines were dumb and incapable of emotion, these creators were confident in staging frank simulations. What interested them were machinic motions evoking drives, instincts, and affects; they mimicked sexual and animal behaviors, as if below the threshold of consciousness. Such artists were uninterested in the manipulation of data or information (although Hans Haacke would move in that direction by 1972 with his "Real-Time Systems" works). The cybernetic culture that artists and scientists were putting in place on two continents embedded the human in the technosphere and seduced perception with the graceful and responsive behaviors of the machinic phylum. "Artificial" and "natural" intertwined in this early cybernetic aesthetic.

But it wouldn't end here. Crucial to the expansion of this uncritical, largely masculine set of cybernetic environments would be a radical, critical cohort of astonishing women artists emerging in the 1990s, fully aware of their predecessors in art and technology but perhaps more inspired by the feminist founders of the 1970 journal *Radical Software* and the cultural blast of Donna Haraway's inspiring 1984 polemic, "A Cyborg Manifesto." The creaky gender theater of Paik and Pask, the innocent creatures of Ihnatowicz and Tsai, were mobilized as savvy, performative, and postmodern, as in Lynn Hershman Leeson's "Dollie Clone Series" (1995–98), consisting of the interactive assemblages *Cybe-Roberta* and *Tillie, the Telerobotic Doll,* who worked the technosphere with the professionalism of burlesque, winking and folding us viewers into an explicit consciousness of our voyeuristic position as both seeing subjects and objects to be looked at.

The "innocent" technosphere established by male cybernetic sculptors of the 1960s was, by the 1990s, identified by feminist artists as an entirely suffusive condition demanding our critical attention. At the same time, feminists tackled the question of whose "intelligence" AI was attempting to simulate. For an artist such as Hershman Leeson, responding to the technical "triumph" of cloning Dolly the sheep, it was crucial to draw the connection between meat production and "meat machines." Hershman Leeson produced "dolls" as clones, offering a critical framing of the way contemporary individuation had become part of an ideological, replicative, plastic realm.

While the technofeminists of the 1990s and into the 2000s weren't all cyber all the time, their works nonetheless complicated the dominant machinic and kinetic qualities of male artists' previous technoenvironments. The androgynous telecyborg in Judith Barry's *Imagination, Dead Imagine* (1991), for example, had no moving parts: He/she was comprised of pure signals, flickering projections on flat surfaces. In her setup, Barry commented on the alienating effects of late-20th-century technology. The image of an androgynous head fills an enormous cube made of ten-foot-square screens on five sides, mounted on a ten-foot-wide mirrored base. A variety of viscous and unpleasant-looking fluids (yellow, reddish-orange, brown), dry materials (sawdust? flour?), and even insects drizzle or dust their way down the head, whose stoic sublimity is made gorgeously virtual on the work's enormous screens. *Dead Imagine,* through its large-scale and cubic "Platonic" form, remains both artificial and locked into the body—refusing a detached "intelligence" as being no intelligence at all.

Artists in the new millennium inherit this critical tradition and inhabit the current paradigms of AI, which has slid from partial simulations to claims of intelligence. In the 1955 proposal thought to be the first printed usage of the phrase "artificial intelligence," computer scientist John McCarthy and his colleagues Marvin Minsky, Nathaniel Rochester, and Claude Shannon conjectured that "every aspect of learning or any other feature of intelligence can in principle be so precisely described that a machine can be made to simulate it." This modest theoretical goal has inflated over the past sixty-four years and is now expressed by Google DeepMind as an ambition to "Solve intelligence." Crack the code! But unfortunately, what we hear cracking is not code but small-scale capitalism, the social contract, and the scaffolding of civility. Taking away the jobs of taxi and truck drivers, roboticizing direct marketing, hegemonizing entertainment, privatizing utilities, and depersonalizing health care—are these the "whips" that Wiener feared we would learn to love?

Artists can't solve any of this. But they can remind us of the creative potential of the paths not taken—the forks in the road that were emerging around 1970, before "information" became capital and "intelligence" equaled data harvesting. Richly evocative of what can be done with contemporary tools when revisiting earlier possibilities is French artist Philippe Parreno's "firefly piece," so nicknamed to avoid having to iterate its actual title: *With a Rhythmic Instinct to Be Able to Travel Beyond Existing Forces of Life* (2014). Described by the artist as "an automaton," the sculptural installation juxtaposes a flickering projection of black-and-white drawings of fireflies with a band of oscillating green-on-black binary figures. The drawings and binary figures are animated using algorithms from mathematician John Horton Conway's 1970 Game of Life, a "cellular automaton."

Conway set up parameters for any square ("cell") to be lit ("alive") or dark ("dead") in an infinite, two-dimensional grid. The rules are summarized as follows: A single cell will quickly die of loneliness. But a cell touching three or more other "live" cells will also die, "due to crowding." A cell survives and thrives if it has just two neighbors . . . and so on. As one cell dies, it may create the conditions for other cells to survive, yielding patterns that appear to move and grow, shifting across the grid like evanescent neural impulses or bioluminescent clusters of diatoms. In Stephen Hawking's 2012 film *The Meaning of Life*, the narrator describes Conway's mathematical model as simulating "how a complex thing like the mind might come about from a basic set of rules," revealing the overweening ambitions that characterize contemporary AI: "[T]hese complex properties emerge from simple laws that contain no concepts like movement or reproduction," yet they produce "species," and cells "can even reproduce, just as life does in the real world."*

Just as life does? Artists know the blandishments of simulation and representation, the difference between the genius of artifice and the realities of what "life does." Parreno's piece is an intuitive assembly of our experience of "life" through embodied, perspectival engagement. Our consciousness is electrically (cybernetically) enmeshed, yet we don't respond as if this human-generated set of elegant simulations had its own *intelligence.*

The artistic use of cybernetic beings also reminds us that consciousness itself is not just "in here." It is streaming in and out, harmonizing those sensory, scintillating signals. Mind happens well outside the limits of the cranium (and its simulacrum, the "motherboard"). In Mary Catherine Bateson's paraphrase of her father, Gregory's, second-order cybernetics, mind is material "not necessarily defined by a boundary such as an envelope of skin."† Parreno pairs the simulations of art with the simulations of mathematics to force the Wiener-like point that any such model is not, by itself, just like life. Models are just that—parts of signaling systems constituting "intelligence" only when their creaturely counterparts engage them in lively meaning making. Contemporary AI has talked itself into a corner by instrumentalizing and particularizing tasks and subroutines, confusing these drills with actual wisdom. The brief cultural history offered here reminds us that views of data as intelligence, digital nets as "neural," or isolated individuals as units of life were alien even to Conway's brute simulation.

We can stigmatize the stubborn arrogance of current AI as "right cybernetics," the path that led to current automated weapons systems, Uber's ill-disguised hostility to human workers, and the capitalist dreams of Google. Now we must turn back to left cybernetics—theoretical biologists and anthropologists engaged with a trans-species understanding of intelligent systems. Gregory Bateson's observation that corporations

---

* Narration in Stephen Hawking's *The Meaning of Life* (Smithson Productions, Discovery Channel, 2012).
† Mary Catherine Bateson, 1999 foreword to Gregory Bateson, *Steps to an Ecology of Mind* (Chicago: University of Chicago Press, 1972), xi.

merely simulate "aggregates of parts of persons," with profit-maximizing decisions cut off from "wider and wiser parts of the mind," has never been more timely.*

The cybernetic epistemology offered here suggests a new approach. The individual mind is immanent, not only in the body but also in pathways outside the body, and there is a larger Mind, of which the individual mind is only a subsystem. This larger Mind, Bateson holds, is comparable to God, and is perhaps what some people mean by "God," but it is still immanent in the total interconnected social system and planetary ecology. This is not the collective delusion of an exterior "God" who speaks from outside human consciousness (this long-seated monotheistic conceit, Bateson suggests, leads to views of nature and environment as also outside the "individual" human, rendering them as "gifts to exploit"). Rather, Bateson's "God" is a placeholder for our evanescent experience of interacting consciousness-in-the-world: larger Mind as a result of inputs and actions that then become inputs for other actions in concert with other entities—webs of symbiotic relationships that form patterns we need urgently to sense and harmonize with.†

From Tsai in the 1970s to Hershman Leeson in the 1990s to Parreno in 2014, artists have been critiquing right cybernetics and plying alternative, embodied, environmental experiences of "artificial" intelligence. Their artistic use of cybernetic beings offers the wisdom of symbionts experienced in the kinds of poiesis that can be achieved in this world: rhythms of signals and intuitive actions that produce the movements of life partnered with an electromechanical and -magnetic technosphere. Life, in its mysterious negentropic entanglements with matter and Mind.

---

* *Steps to an Ecology of Mind*, 452.
† *Steps to an Ecology of Mind*, 467–68.

Chapter 25

ARTIFICIAL INTELLIGENCE AND THE FUTURE OF CIVILIZATION

STEPHEN WOLFRAM

Stephen Wolfram is a scientist, inventor, and the founder and CEO of Wolfram Research. He is the creator of the symbolic computation program Mathematica and its programming language, Wolfram Language, as well as the knowledge engine Wolfram\Alpha. He is also the author of A New Kind of Science. The following is an edited transcript from a live interview with him conducted in December 2015.

Over nearly four decades, **Stephen Wolfram** has been a pioneer in the development and application of computational thinking and responsible for many innovations in science, technology, and business.

His 1982 paper "Cellular Automata as Simple Self-Organizing Systems," written at the age of twenty-three, was the first of numerous significant scientific contributions aimed at understanding the origins of complexity in nature.

It was around this time that Stephen briefly came into my life. I had established The Reality Club, an informal gathering of intellectuals who met in New York City to present their work before peers in other disciplines. (Note: In 1996, The Reality Club went online as Edge.org.) Our first speaker? Stephen Wolfram, a "wunderkind" who had arrived in Princeton at the Institute for Advanced Study. I distinctly recall his focused manner as he sat down on a couch in my living room and spoke uninterrupted for about an hour before the assembled group.

Since that time, Stephen has become intent on making the world's knowledge easily computable and accessible. His program Mathematica is the definitive system for modern technical computing. Wolfram|Alpha computes expert-level answers using AI technology. He considers his Wolfram Language to be the first true computational communication language for humans and AIs.

I caught up with him again four years ago, when we arranged to meet in Cambridge, Massachusetts, for a freewheeling conversation about AI. Stephen walked in, said hello, sat down, and, looking at the video camera set up to record the conversation for Edge, began to talk and didn't stop for two and a half hours.

The essay that follows is an edited version of that session, which was a Wolfram master class of sorts and is an appropriate way to end this volume—just as Stephen's Reality Club talk in the eighties was a great way to initiate the ongoing intellectual enterprise whose result is the rich community of thinkers presenting their work to one another and to the public in this book.

see technology as taking human goals and making them automatically executable by machines. Human goals of the past have entailed moving objects from here to there, using a forklift rather than our own hands. Now the work we can do automatically, with machines, is mental rather than physical. It's obvious that we can automate many of the tasks we humans have long been proud of doing ourselves. What's the future of the human condition in that situation?

People talk about the future of intelligent machines and whether they'll take over and decide what to do for themselves. But the inventing of goals is not something that has a path to automation. Someone or something has to define what a machine's purpose should be—what it's trying to execute. How are goals defined? For a given human, they tend to be defined by personal history, cultural environment, the history of our civilization. Goals are uniquely human. Where the machine is concerned, we can give it a goal when we build it.

What kinds of things have intelligence, or goals, or purpose? Right now, we know one great example, and that's us—our brains, our human intelligence. Human intelligence, I once assumed, is far beyond anything else that exists naturally in the world; it's the result of an elaborate process of evolution and thus stands apart from the rest of existence. But what I've realized, as a result of the science I've done, is that this is not the case.

People might say, for instance, "The weather has a mind of its own."

That's an animist statement and seems to have no place in modern scientific thinking. But it's not as silly as it sounds. What does the human brain do? A brain receives certain input, it computes things, it causes certain actions to happen, it generates a certain output. Like the weather. All sorts of systems are, effectively, doing computations—whether it's a brain or, say, a cloud responding to its thermal environment.

We can argue that our brains are doing vastly more sophisticated computations than those in the atmosphere. But it turns out that there's a broad equivalence between the kinds of computations that different kinds of systems do. This renders the question of the human condition somewhat poignant, because it seems we're not as special as we thought. There are all those different systems of nature that are pretty much equivalent, in terms of their computational capabilities.

What makes us different from all those other systems is the particulars of our history, which give us our notions of purpose and goals. That's a long way of saying that when the box on our desk thinks as well as the human brain does, what it still won't have, intrinsically, are goals and purposes. Those are defined by our particulars—our particular biology, our particular psychology, our particular cultural history.

When we consider the future of AI, we need to think about the goals. That's what humans contribute; that's what our civilization contributes. The execution of those goals is what we can increasingly automate. What will the future of humans be in such a world? What will there be for them to do? One of my projects has been to understand the evolution of human purposes over time. Today we've got all kinds of purposes. If you look back a thousand years, people's goals were quite different: How do I get my food? How do I keep myself safe? In the modern Western world, for the most part you don't spend a large fraction of your life thinking about those purposes. From the point of view of a thousand years ago, some of the goals people have today would seem utterly bizarre—for example, like exercising on a treadmill. A thousand years ago that would sound like a crazy thing to do.

What will people be doing in the future? A lot of purposes we have today are generated by scarcity of one kind or another. There are scarce resources in the world. People want to get more of something. Time itself is scarce in our lives. Eventually, those forms of scarcity will disappear. The most dramatic discontinuity will surely be when we achieve effective human immortality. Whether this will be achieved biologically or digitally isn't clear, but inevitably it will be achieved. Many of our current goals are driven in part by our mortality: "I'm only going to live a certain time, so I'd better get this or that done." And what happens when most of our goals are executed automatically? We won't have the kinds of motivations we have today. One question I'd like an answer for is, What do the derivatives of humans in the future end up choosing to do with themselves? One of the potential bad outcomes is that they just play video games all the time.

The term "artificial intelligence" is evolving in its use in technical language. These days, AI is very popular, and people have some idea of what it means. Back when computers were being developed, in the 1940s and 1950s, the typical title of a book or a magazine article about computers was "Giant Electronic Brains." The idea was that just as bulldozers and steam engines and so on automated mechanical work, computers would automate intellectual work. That promise turned out to be harder to fulfill than many people expected. There was, at first, a great deal of optimism; a lot of government money got spent on such efforts in the early 1960s. They basically just didn't work.

There are a lot of amusing science-fiction-ish portrayals of computers in the movies of that time. There's a cute one called *Desk Set*, which is about an IBM-type computer being installed in a broadcasting company and putting everybody out of a job. It's cute because the computer gets asked a bunch of reference-library questions. When my colleagues and I were building Wolfram|Alpha, one of the ideas we had was to get it to answer all of those reference-library questions from *Desk Set*. By 2009, it could answer them all.

In 1943, Warren McCulloch and Walter Pitts came up with a model for how brains conceptually, formally, might work—an artificial neural network. They saw that their brainlike model would do computations in the same way as Turing Machines. From their work, it emerged that we could make brainlike neural networks that would act as general computers. And in fact, the practical work done by the ENIAC folks and John von Neumann and others on computers came directly not from Turing Machines but through this bypath of neural networks.

But simple neural networks didn't do much. Frank Rosenblatt invented a learning device he called the perceptron, which was a one-layer neural network. In the late sixties, Marvin Minsky and Seymour Papert wrote a book titled *Perceptrons*, in which they basically proved that perceptrons couldn't do anything interesting, which is correct. Perceptrons could make only linear distinctions between things. So the idea was more or less dropped. People said, "These guys have written a proof that neural networks can't do anything interesting, therefore no neural networks can do anything interesting, so let's forget about neural networks." That attitude persisted for some time.

Meanwhile, there were a couple of other approaches to AI. One was based on understanding, at a formal level, symbolically, how the world works; and the other was based on doing statistics and probabilistic kinds of things. With regard to symbolic AI, one of the test cases was, Can we teach a computer to do something like integrals? Can we teach a computer to do calculus? There were tasks like machine translation, which people thought would be a good example of what computers could do. The bottom line is that by the early seventies, that approach had crashed.

Then there was a trend toward devices called expert systems, which arose in the late seventies and early eighties. The idea was to have a machine learn the rules that an expert uses and thereby figure out what to do. That petered out. After that, AI became little more than a crazy pursuit.

I had been interested in how you make an AI-like machine since I was a kid. I was interested particularly in how you take the knowledge we humans have accumulated in our civilization and automate answering questions on the basis of that knowledge. I thought about how you could do that symbolically, by building a system that could break down questions into symbolic units and answer them. I worked on neural networks at that time and didn't make much progress, so I put it aside for a while.

Back in mid-2002 to 2003, I thought about that question again: What does it take to make a computational knowledge system? The work I'd done by then pretty much showed that my original belief about how to do this was completely wrong. My original belief had been that in order to make a serious computational knowledge system, you first had to build a brainlike device and then feed it knowledge—just as humans learn in standard education. Now I realized that there wasn't a bright line between what is intelligent and what is simply computational.

I had assumed that there was some magic mechanism that made us vastly more capable than anything that was just computational. But that assumption was wrong. This insight is what led to Wolfram|Alpha. What I discovered is that you can take a large collection of the world's knowledge and automatically answer questions on the basis of it, using what are essentially merely computational techniques. It was an alternative way to do engineering—a way that's much more analogous to what biology does in evolution.

In effect, what you normally do when you build a program is build it step-by-step. But you can also explore the computational universe and mine technology from that universe. Typically, the challenge is the same as in physical mining: That is, you find a supply of, let's say, iron, or cobalt, or gadolinium, with some special magnetic properties, and you turn that special capability to a human purpose, to something you want technology to do. In the case of magnetic materials, there are plenty of ways to do that. In terms of programs, it's the same story. There are all kinds of programs out there, even tiny programs that do complicated things. Could we entrain them for some useful human purpose?

And how do you get AIs to execute your goals? One answer is to just talk to them, in the natural language of human utterances. It works pretty well when you're talking to Siri. But when you want to say something longer and more complicated, it doesn't work well. You need a computer language that can represent sophisticated concepts in a way that can be progressively built up and isn't possible in natural language. What my company spent a lot of time doing was building a knowledge-based language that incorporates the knowledge of the world directly into the language. The traditional approach to creating a computer language is to make a language that represents operations that computers intrinsically know how to do: allocating memory, setting values of variables, iterating things, changing program counters, and so on. Fundamentally, you're telling computers to do things in your own terms. My approach was to make a language that panders not to the computers but to the humans, to take whatever a human thinks of and convert it into some form that the computer can understand. Could we encapsulate the knowledge we'd accumulated, both in science and in data collection, into a language we could use to communicate with computers? That's the big achievement of my last thirty years or so—being able to do that.

Back in the 1960s, people would say things like, "When we can do such-and-such, we'll know we have AI"; "When we can do an integral from a calculus course, we'll know we have AI"; "When we can have a conversation with a computer and make it seem human . . ."; etc. The difficulty was, "Well, gosh, the computer just doesn't know enough about the world." You'd ask the computer what day of the week it was, and it might be able to answer that. You'd ask it who the president was, and it probably couldn't tell you. At that point, you'd know you were talking to a computer and not a person. But now when it comes to these Turing Tests, people who've tried connecting, for example, Wolfram|Alpha to their Turing Test bots find that the bots lose every time. Because all you have to do is start asking the machine sophisticated questions and it will answer them! No human can do that. By the time you've asked it a few disparate questions, there will be no human who knows all those things, yet the system will know them. In that sense, we've already achieved good AI, at that level.

Then there are certain kinds of tasks that are easy for humans but traditionally very hard for machines. The standard one is visual object identification: What is this object? Humans can recognize it and give some simple description of it, but a computer was just hopeless at that. A couple of years ago, though, we brought out a little image-identification system, and many other companies have done something similar—ours happens to be somewhat better than the rest. You show it an image, and for about ten thousand kinds of things, it will tell you what it is. It's fun to show it an abstract painting and see what it says. But it does a pretty good job.

It works using the same neural-network technology that McCulloch and Pitts imagined in 1943 and lots of us worked on in the early eighties. Back in the 1980s, people successfully did OCR—optical character recognition. They took the twenty-six letters of the alphabet and said, "OK, is that an A? Is that a B? Is that a C?" and so on. That could be done for twenty-six different possibilities, but it couldn't be done for ten thousand. It was just a matter of scaling up the whole system that makes this possible today. There are maybe five thousand picturable common nouns in English, ten thousand if you include things like special kinds of plants and beetles that people would recognize with some frequency. What we did was train our system on 30 million images of these kinds of things. It's a big, complicated, messy neural network. The details of the network probably don't matter, but it takes about a quadrillion GPU operations to do the training.

Our system is impressive because it pretty much matches what humans can do. It has about the same training data humans have—about the same number of images a human infant would see in the first couple of years of its life. Roughly the same number of operations have to be done in the learning process, using about the same number of neurons in at least the first levels of our visual cortex. The details are different; the way these artificial neurons work has little to do with how the brain's neurons work. But the concept is similar, and there's a certain universality to what's going on. At the mathematical level, it's a composition of a very large number of functions, with certain continuity properties that let you use calculus methods to incrementally train the system. Given those attributes, you can end up with something that does the same job human brains do in physiological recognition.

But does this constitute AI? There are a few basic components. There's physiological recognition, there's voice-to-text, there's language translation—things humans manage to do with varying degrees of difficulty. These are essentially some of the links to how we make machines that are humanlike in what they do. For me, one of the interesting things has been incorporating those capabilities into a precise symbolic language to represent the everyday world. We now have a system that can say, "This is a glass of water." We can go from a picture of a glass of water to the *concept* of a glass of water. Now we have to invent some actual symbolic language to represent those concepts.

I began by trying to represent mathematical, technical kinds of knowledge and went on to other kinds of knowledge. We've done a pretty good job of representing objective knowledge in the world. Now the problem is to represent everyday human discourse in a precise symbolic way—a knowledge-based language intended for communication between humans and machines, so that humans can read it and machines can understand it, too. For instance, you might say, "X is greater than 5." That's a predicate. You might also say, "I want a piece of chocolate." That's also a predicate. It has an "I want" in it. We have to find a precise symbolic representation of the desires we express in human natural language.

In the late 1600s, Gottfried Leibniz, John Wilkins, and others were concerned with what they called philosophical languages—that is, complete, universal, symbolic representations of things in the world. You can look at the philosophical language of John Wilkins and see how he divided up what was important in the world at the time. Some aspects of the human condition have been the same since the 1600s. Some are very different. His section on death and various forms of human suffering was huge; in today's ontology, it's a lot smaller. It's interesting to see how a philosophical language of today would differ from a philosophical language of the mid-1600s. It's a measure of our progress. Many such attempts at formalization have happened over the years. In mathematics, for example: Whitehead and Russell's *Principia Mathematica* in 1910 was the biggest showoff effort. There were previous attempts by Gottlob

Frege and Giuseppe Peano that were a little more modest in their presentation. Ultimately, they were wrong in what they thought they should formalize: They thought they should formalize some process of mathematical proof, which turns out not to be what most people care about.

With regard to a modern analog of the Turing Test, it's an interesting question. There's still the conversational bot, which is Turing's idea. That one hasn't been solved yet. It will be solved—the only question is, What is the application for which it is solved? For a long time I would ask, Why should we care?—because I thought the principal application would be customer service, which wasn't particularly high on my list. But customer service, where you're trying to interface, is just where you need this conversational language.

One big difference between Turing's time and ours is the method of communicating with computers. In his time, you typed something into the machine and it typed back a response. In today's world, it responds with a screen—as, for instance, when you want to buy a movie ticket. How is a transaction with a machine different from a transaction with a human? The main answer is that there's a visual display. It asks you something, and you press a button, and you can see the result immediately. For example, in Wolfram|Alpha, when it's used inside Siri, if there's a short answer, Siri will tell you the short answer. But what most people want is the visual display, showing the infographic of this or that. This is a nonhuman form of communication that turns out to be richer than the traditional spoken, or typed, human communication. In most human-to-human communication, we're stuck with pure language, whereas in computer-to-human communication we have this much higher bandwidth channel—of visual communication.

Many of the most powerful applications of the Turing Test fall away now that we have this additional communication channel. For example, here's one we're pursuing right now. It's a bot that communicates about writing programs: You say, "I want to write a program. I want it to do this." The bot will say, "I've written this piece of program. This is what it does. Is this what you want?" Blah-blah-blah. It's a back-and-forth bot. Devising such systems is an interesting problem, because they have to have a model of a human if they're trying to explain something to you. They have to know what the human is confused about.

What has long been difficult for me to understand is, What's the point of a conventional Turing Test? What's the motivation? As a toy, one could make a little chat bot that people could chat with. That will be the next thing. The current round of deep learning—particularly, recurrent neural networks—is making pretty good models of human speech and human writing. We can type in, say, "How are you feeling today?" and it knows most of the time what sort of response to give. But I want to figure out whether I can automate responding to my email. I know the answer is no. A good Turing Test, for me, will be when a bot can answer most of my email. That's a tough test. It would have to learn those answers from the human the email is connected to. I might be a little bit ahead of the game, because I've been collecting data on myself for about twenty-five years. I have every piece of email for twenty-five years, every keystroke for twenty. I should be able to train an avatar, an AI, that will do what I can do—perhaps better than I could.

People worry about the scenario in which AIs take over. I think something much more amusing, in a sense, will happen first. The AI will know what you intend, and it will be good at figuring out how to get there. I tell my car's GPS I want to go to a particular destination. I don't know where the heck I am, I just follow my GPS. My children like to remind me that once when I had a very early GPS—the kind that told you, "Turn this way, turn that way"—we ended up on one of the piers going out into Boston Harbor.

More to the point is that there will be an AI that knows your history, and knows that when you're ordering dinner online you'll probably want such-and-such, or when you email this person, you should talk to them about such-and-such. More and more, the AIs will suggest to us what we should do, and I suspect most of the time people will just go along with that. It's good advice—better than what you would have figured out for yourself.

As far as the takeover scenario is concerned, you can do terrible things with technology and you can do good things with technology. Some people will try to do terrible things with technology, and some people will try to do good things with technology. One of the things I like about today's technology is the equalization it has produced. I used to be proud that I had a better computer than anybody I knew; now we all have the same kinds of computers. We have the same smartphones, and pretty much the same technology can be used by a decent fraction of the planet's 7 billion people. It's not the case that the king's technology is different from everybody else's. That's an important advance.

The great frontier five hundred years ago was literacy. Today, it's doing programming of some kind. Today's programming will be obsolete in a not very long time. For example, people no longer learn assembly language, because computers are better at writing assembly language than humans are, and only a small set of people need to know the details of how language gets compiled into assembly language. A lot of what's being done by armies of programmers today is similarly mundane. There's no good reason for humans to be writing Java code or JavaScript code. We want to automate the programming process so that what's important goes from what the human wants done to getting the machine, as automatically as possible, to do it. This will increase that equalization, which is something I'm interested in. In the past, if you wanted to write a serious piece of code, or program for something important and real, it was a lot of work. You had to know quite a bit about software engineering, you had to invest months of time in it, you had to hire programmers who knew this or you had to learn it yourself. It was a big investment.

That's not true anymore. A one-line piece of code already does something interesting and useful. It allows a vast range of people who couldn't make computers do things for them make computers do things for them. Something I'd like to see is a lot of kids around the world learn the new capabilities of knowledge-based programming and then produce code that's effectively as sophisticated as what anybody in the top ranks can produce. This is within reach. We're at the point where anybody can learn to do knowledge-based programming, and, more important, learn to think computationally. The actual mechanics of programming are easy now. What's difficult is imagining things in a computational way.

How do you teach computational thinking? In terms of how to do programming, it's an interesting question. Take nanotechnology. How did we achieve nanotechnology? Answer: We took technology as we understand it on a large scale and we made it very small. How to make a CPU chip on the atomic scale? Fundamentally, we use the same architecture as the CPU chip we know and love. That isn't the only approach one can take. Looking at what simple programs can do suggests that you can take even simple impoverished components and, with the right compiler, make them do interesting things. We don't do molecular-scale computing yet, because the ambient technology is such that you'd have to spend a decade building it. But we've got the components that are enough to make a universal computer. You might not know how to program with those components, but by doing searches in the space of possible programs, you'd start to amass building blocks, and you could then create a compiler for them. The surprising thing is that impoverished stuff is capable of doing sophisticated things, and the compilation step is not as gruesome as you might expect.

Just searching the computational universe and trying to find programs—building blocks—that are interesting is a good approach. A more traditional engineering approach—trying by pure thought to figure out how to build a universal computer—is a harder row to hoe. That doesn't mean it can't be done, but my guess is that we'll be able to do some amazing things just by finding the components and searching the possible programs we can make with them. Then it's back to the question about connecting human purposes to what is available from the system.

One question I'm interested in is, What will the world look like when most people can write code? We had a transition, maybe five hundred or so years ago, when only scribes and a small set of the population could read and write natural language. Today, a small fraction of the population can write code. Most of the code they write is for computers only. You don't understand things by reading code. But there will come a time when, as a result of things I've tried to do, the code is at a high enough level that it's a minimal description of what you're trying to do. It will be a piece of code that's understandable to humans but also executable by the machines.

Coding is a form of expression, just as writing in a natural language is a form of expression. To me, some simple pieces of code are poetic—they express ideas in a very clean way. There's an aesthetic aspect, much as there is to expression in a natural language. One feature of code is that it's immediately executable; it's not like writing. When you write something, somebody has to read it, and the brain that's reading it has to absorb the thoughts that came from the person who did the writing. Look at how knowledge has been transmitted in the history of the world. At level zero, one form of knowledge transmission is essentially genetic—that is, there's an organism, and its progeny has the same features that it has. Then there's the kind of knowledge transmission that happens with things like physiological recognition. A newborn creature has some neural network with some random connections in it, and as the creature moves around in the world, it starts recognizing kinds of objects and it learns that knowledge.

Then there's the level that was the big achievement of our species, which is natural language: the ability to represent knowledge abstractly enough that we can communicate it brain to brain, so to speak. Arguably, natural language is our species' most important invention. It's what led, in many respects, to our civilization.

There's yet another level, and probably one day it will have a more interesting name. With knowledge-based programming, we have a way of creating an actual representation of real things in the world in a precise and symbolic way. Not only is it understandable by brains and communicable to other brains and to computers, it's also immediately executable.

Just as natural language gave us civilization, knowledge-based programming will give us—what? One bad answer is that it will give us the civilization of the AIs. That's what we don't want to happen, because the AIs will do a great job communicating with one another and we'll be left out of it, because there's no intermediate language, no interface with our brains. What will this fourth level of knowledge communication lead to? If you were Caveman Ogg and you were just realizing that language was starting, could you imagine the coming of civilization? What should we be imagining right now?

This relates to the question of what the world would look like if most people could code. Clearly, many trivial things would change: Contracts would be written in code, restaurant recipes might be written in code, and so on. Simple things like that would change. But much more profound things would also change. The rise of literacy gave us bureaucracy, for example, which had already existed but dramatically accelerated, giving us a greater depth of governmental systems, for better or worse. How does the coding world relate to the cultural world?

Take high school education. If we have computational thinking, how does that affect how we study history? How does that affect how we study languages, social studies, and so on? The answer is, it has a great effect. Imagine you're writing an essay. Today, the raw material for a typical high school student's essay is something that's already been written; students usually can't generate new knowledge easily. But in the computational world, that will no longer be true. If the students know something about writing code, they'll access all that digitized historical data and figure out something new. Then they'll write an essay about something they've discovered. The achievement of knowledge-based programming is that it's no longer sterile, because it's got the knowledge of the world knitted into the language you're using to write code.

There's computation all over the universe: in a turbulent fluid producing some complicated pattern of flow, in the celestial mechanics of planetary interactions, in brains. But does computation have a purpose? You can ask that about any system. Does the weather have a goal? Does climate have a goal?

Can someone looking at Earth from space tell that there's anything with a purpose there? Is there a civilization there? In the Great Salt

Lake in Utah there's a straight line. It turns out to be a causeway dividing two areas of the lake with different colors of algae, so it's a very dramatic straight line. There's a road in Australia that's long and straight. There's a railroad in Siberia that's long, and lights go on when a train stops at the stations. So from space you can see straight lines and patterns.

But are these clear enough examples of obvious purpose on Earth as viewed from space? For that matter, how do we recognize extraterrestrials out there? How do we tell if a signal we're getting indicates purpose? Pulsars were discovered in 1967, when we picked up a periodic flutter every second or so. The first question was, Is this a beacon? Because what else would make a periodic signal? It turned out to be a rotating neutron star.

One criterion to apply to a potentially purposeful phenomenon is whether it's minimal in achieving a purpose. But does that mean that it was built for the purpose? The ball rolls down the hill because of gravitational pull. Or the ball rolls down the hill because it's satisfying the principle of least action. There are typically these two explanations for some action that seems purposeful: the mechanistic explanation and the teleological. Essentially all of our existing technology fails the test of being minimal in achieving its purpose. Most of what we build is steeped in technological history, and it's incredibly nonminimal for achieving its purpose. Look at a CPU chip; there's no way that that's the minimal way to achieve what a CPU chip achieves.

This question of how to identify purposefulness is a hard one. It's an important question, because radio noise from the galaxy is very similar to CDMA transmissions from cell phones. Those transmissions use pseudonoise sequences, which happen to have certain repeatability properties. But they come across as noise, and they're set up as noise, so as not to interfere with other channels. The issue gets messier. If we were to observe a sequence of primes being generated from a pulsar, we'd ask what generated them. Would it mean that a whole civilization grew up and discovered primes and invented computers and radio transmitters and did this? Or is there just some physical process making primes? There's a little cellular automaton that makes primes. You can see how it works if you take it apart. It has a little thing bouncing inside it, and out comes a sequence of primes. It didn't need the whole history of civilization and biology and so on to get to that point.

I don't think there is abstract "purpose," per se. I don't think there's abstract meaning. Does the universe have a purpose? Then you're doing theology in some way. There is no meaningful sense in which there is an abstract notion of purpose. Purpose is something that comes from history.

One of the things that might be true about our world is that maybe we go through all this history and biology and civilization, and at the end of the day the answer is "42," or something. We went through all those 4 billion years of various kinds of evolution and then we got to "42."

Nothing like that will happen, because of computational irreducibility. There are computational processes that you can go through in which there is no way to shortcut that process. Much of science has been about shortcutting computation done by nature. For example, if we're doing celestial mechanics and want to predict where the planets will be a million years from now, we could follow the equations, step-by-step. But the big achievement in science is that we're able to shortcut that and reduce the computation. We can be smarter than the universe and predict the endpoint without going through all the steps. But even with a smart enough machine and smart enough mathematics, we can't get to the endpoint without going through the steps. Some details are irreducible. We have to irreducibly follow those steps. That's why history means something. If we could get to the endpoint without going through the steps, history would be, in some sense, pointless.

So it's not the case that we're intelligent and everything else in the world is not. There's no enormous abstract difference between us and the clouds or us and the cellular automata. We cannot say that this brainlike neural network is qualitatively different from this cellular-automaton system. The difference is a detailed difference. This brainlike neural network was produced by the long history of civilization, whereas the cellular automaton was created by my computer in the last microsecond.

The problem of abstract AI is similar to the problem of recognizing extraterrestrial intelligence: How do you determine whether or not it has a purpose? This is a question I don't consider answered. We'll say things like, "Well, AI will be intelligent when it can do blah-blah-blah." When it can find primes. When it can produce this and that and the other. But there are many other ways to get to those results. Again, there is no bright line between intelligence and mere computation.

It's another part of the Copernican story: We used to think Earth was the center of the universe. Now we think we're special because we have intelligence and nothing else does. I'm afraid the bad news is that that isn't a distinction.

Here's one of my scenarios. Let's say there comes a time when human consciousness is readily uploadable into digital form, virtualized and so on, and pretty soon we have a box of a trillion souls. There are a trillion souls in the box, all virtualized. In the box, there will be molecular computing going on—maybe derived from biology, maybe not. But the box will be doing all kinds of elaborate stuff. And there's a rock sitting next to the box. Inside a rock, there is always all kinds of elaborate stuff going on, all kinds of subatomic particles doing all kinds of things. What's the difference between the rock and the box of a trillion souls? The answer is that the details of what's happening in the box were derived from the long history of human civilization, including whatever people watched on YouTube the day before. Whereas the rock has its long geological history but not the particular history of our civilization.

Realizing that there isn't a genuine distinction between intelligence and mere computation leads you to imagine that future—the endpoint of our civilization as a box of a trillion souls, each of them essentially playing a video game, forever. What is the "purpose" of that?

Index